全国普通高等学校机械类"十二五"规划系列教材

现代机械工程专业英语
English in Modern Mechanical Engineering

主　编　赵运才　李秀辰
副主编　黄丽蓉　王　静　林秋实
主　审　严宏志

华中科技大学出版社
中国·武汉

内 容 简 介

本书按照"从机械设计制造及自动化、汽车工程专业出发,注意现代机电技术的发展,将机、电、液相结合"的整体原则组织编写。全书分为三大部分:机械基础英语、现代机械装备专业英语和专业英语应用。内容涉及力学、机械零件与机构、机械设计、汽车构造和工作性能、机械加工及成形技术、现代机械装备等方面知识。特别在书中编写了环保装备、再制造工程、可持续产品设计和节能装备等与当前现代经济社会发展密切相关的内容。全书共有26篇课文,并附有参考译文。

本书可以作为机械设计及自动化、汽车工程、机电工程等专业的专业英语教材,也可以供从事相关专业工作的科技人员参考使用。

图书在版编目(CIP)数据

现代机械工程专业英语/赵运才　李秀辰　主编.—武汉:华中科技大学出版社,2013.8(2023.3重印)
ISBN 978-7-5609-8542-8

Ⅰ.现… Ⅱ.①赵… ②李… Ⅲ.机械工程-英语-高等学校-教材　Ⅳ.H31

中国版本图书馆 CIP 数据核字(2012)第 276292 号

现代机械工程专业英语　　　　　　　　　　　　　　　　　　　　赵运才　李秀辰　主编

策划编辑:俞道凯
责任编辑:姚同梅
封面设计:范翠璇
责任校对:张　琳
责任监印:张正林
出版发行:华中科技大学出版社(中国·武汉)　　电话:(027)81321913
　　　　　武汉市东湖新技术开发区华工科技园　　邮编:430223
录　　排:华中科技大学惠友文印中心
印　　刷:武汉邮科印务有限公司
开　　本:787mm×1092mm　1/16
印　　张:18.5
字　　数:490千字
版　　次:2023年3月第1版第8次印刷
定　　价:49.80元

本书若有印装质量问题,请向出版社营销中心调换
全国免费服务热线:400-6679-118　　竭诚为您服务
版权所有　侵权必究

全国普通高等学校机械类"十二五"规划系列教材

编审委员会

顾　问：李培根　华中科技大学
　　　　林萍华　华中科技大学

主　任：吴昌林　华中科技大学

副主任：（按姓氏笔画顺序排列）
　　　　王生武　邓效忠　轧　钢　庄哲峰　吴　波　何岭松
　　　　陈　炜　杨家军　杨　萍　竺志超　高中庸　谢　军

委　员：（排名不分先后）
　　　　许良元　程荣龙　曹建国　郭克希　朱贤华　贾卫平
　　　　丁晓非　张生芳　董　欣　庄哲峰　蔡业彬　许泽银
　　　　许德璋　叶大鹏　李耀刚　耿　铁　邓效忠　宫爱红
　　　　成经平　刘　政　王连弟　张庐陵　张建国　郭润兰
　　　　张永贵　胡世军　汪建新　李　岚　杨术明　杨树川
　　　　李长河　马晓丽　刘小健　汤学华　孙恒五　聂秋根
　　　　赵　坚　马　光　梅顺齐　蔡安江　刘俊卿　龚曙光
　　　　吴凤和　李　忠　罗国富　张　鹏　张禹君　柴保明
　　　　孙　未　何　庆　李　理　孙文磊　李文星　杨咸启

秘　书：俞道凯　万亚军

全国普通高等学校机械类"十二五"规划系列教材

序

"十二五"时期是全面建设小康社会的关键时期,是深化改革开放、加快转变经济发展方式的攻坚时期,也是贯彻落实《国家中长期教育改革和发展规划纲要(2010—2020年)》的关键五年。教育改革与发展面临着前所未有的机遇和挑战。以加快转变经济发展方式为主线,推进经济结构战略性调整、建立现代产业体系,推进资源节约型、环境友好型社会建设,迫切需要进一步提高劳动者素质,调整人才培养结构,增加应用型、技能型、复合型人才的供给。同时,当今世界处在大发展、大调整、大变革时期,为了迎接日益加剧的全球人才、科技和教育竞争,迫切需要全面提高教育质量,加快拔尖创新人才的培养,提高高等学校的自主创新能力,推动"中国制造"向"中国创造"转变。

为此,近年来教育部先后印发了《教育部关于实施卓越工程师教育培养计划的若干意见》(教高[2011]1号)、《关于"十二五"普通高等教育本科教材建设的若干意见》(教高[2011]5号)、《关于"十二五"期间实施"高等学校本科教学质量与教学改革工程"的意见》(教高[2011]6号)、《教育部关于全面提高高等教育质量的若干意见》(教高[2012]4号)等指导性意见,对全国高校本科教学改革和发展方向提出了明确的要求。在上述大背景下,教育部高等学校机械学科教学指导委员会根据教育部高教司的统一部署,先后起草了《普通高等学校本科专业目录机械类专业教学规范》、《高等学校本科机械基础课程教学基本要求》,加强教学内容和课程体系改革的研究,对高校机械类专业和课程教学进行指导。

为了贯彻落实教育规划纲要和教育部文件精神,满足各高校高素质应用型高级专门人才培养要求,根据《关于"十二五"普通高等教育本科教材建设的若干意见》文件精神,华中科技大学出版社在教育部高等学校机械学科教学指导委员会的指导下,联合一批机械学科办学实力强的高等学校、部分机械特色专业突出的学校和教学指导委员会委员、国家级教学团队负责人、国家级教学名师组成编委会,邀请来自全国高校机械学科教学一线的教师组织编写全国普通高等学校机械类"十二五"规划系列教材,将为提高高等教育本科教学质量和人才培养质量提供有力保障。

当前,经济社会的发展,对高校的人才培养质量提出了更高的要求。该套教材在编写中,应着力构建满足机械工程师后备人才培养要求的教材体系,以机械工程知识和能力的培养为根本,与企业对机械工程师的能力目标紧密结合,力求满足学科、教学和社会三方面的需求;在结构上和内容上体现思想性、科学性、先进性,把握行业人才要求,突出工程教育特色。同时,注意吸收教学指导委员会教学内容和课程体系改革的研究成果,根据教指委颁布的各课程教学专业规范要求编写,开发教材配套资源(习题、课程设计和实践教材及数字化学习资源),适应新时期教学需要。

教材建设是高校教学中的基础性工作,是一项长期的工作,需要不断吸取人才培养模式和教学改革成果,吸取学科和行业的新知识、新技术、新成果。本套教材的编写出版只是近年来各参与学校教学改革的初步总结,还需要各位专家、同行提出宝贵意见,以进一步修订、完善,不断提高教材质量。

谨为之序。

国家级教学名师
华中科技大学教授、博导
2012 年 8 月

前　言

　　机电工程专业英语是机械设计制造及自动化、汽车工程等专业的一门重要的基础课。对于机电工程专业的本科、专科学生以及从事相关专业工作的科技人员来说，熟练掌握专业英语对于促进国际交流、了解国内外本专业的最新发展动态是十分必要的，并且有着越来越重要的意义。随着我国与国外的技术交流越来越多，对专业英语知识的学习需求更为迫切。为了满足机械设计制造及自动化、汽车工程专业教学需求，我们编写了《现代机械工程专业英语》一书。

　　本书内容由浅入深、由简到繁、循序渐进，同时结合当前现代经济社会发展选材，内容丰富，语言规范，难度适中，便于自学。通过这本教材，学生们不仅可以熟练和掌握本专业常用的及本专业相关的英语单词、词组及其用法，而且可以深化对本专业知识的理解和了解最新的相关专业知识，从而为今后的学习和工作打下良好的基础。

　　本书由江西理工大学赵运才、大连海洋大学李秀辰任主编，江西理工大学黄丽蓉、青岛理工大学王静、昆明学院林秋实任副主编，参加编写的有江西农业大学杨红飞、吉林农业大学杨丹、白城师范学院李温温、内蒙古民族大学张丹丹（排序不分先后），由中南大学机电工程学院严宏志担任主审。由于水平有限，书中错误和不当之处在所难免，欢迎广大读者不吝指教。

编　者

2012 年 7 月

目 录

Lesson 1　Engineering Drawings ································· (1)
　Text ··· (1)
　New Words and Expressions ································ (4)
　Notes ··· (5)
　Translation Skills ··· (5)
　Reading Material ·· (9)
　New Words and Expressions ································ (11)

Lesson 2　Mechanics ·· (12)
　Text ··· (12)
　New Words and Expressions ································ (14)
　Notes ··· (14)
　Translation Skills ··· (15)
　Reading Material ·· (18)
　New Words and Expressions ································ (19)

Lesson 3　Engineering Materials ································ (20)
　Text ··· (20)
　New Words and Expressions ································ (22)
　Notes ··· (22)
　Translation Skills ··· (22)
　Reading Material ·· (25)
　New Words and Expressions ································ (27)

Lesson 4　Mechanical Design ···································· (28)
　Text ··· (28)
　New Words and Expressions ································ (31)
　Notes ··· (31)
　Translation Skills ··· (32)
　Reading Material ·· (34)
　New Words and Expressions ································ (35)

Lesson 5　Machinery Components ····························· (36)
　Text ··· (36)
　New Words and Expressions ································ (40)
　Notes ··· (40)
　Translation Skills ··· (41)
　Reading Material ·· (43)
　New Words and Expressions ································ (46)

Lesson 6　The Basic Components of an Automobile	(47)
Text	(47)
New Words and Expressions	(49)
Notes	(50)
Translation Skills	(50)
Reading Material	(53)
New Words and Expressions	(55)
Lesson 7　Manufacturing Process	(56)
Text	(56)
New Words and Expressions	(58)
Notes	(58)
Translation Skills	(59)
Reading Material	(62)
New Words and Expressions	(64)
Lesson 8　Tolerance and Interchangeability	(66)
Text	(66)
New Words and Expressions	(69)
Notes	(70)
Translation Skills	(70)
Reading Material	(72)
New Words and Expressions	(73)
Lesson 9　Numerical Control	(74)
Text	(74)
New Words and Expressions	(76)
Notes	(77)
Translation Skills	(77)
Reading Material	(80)
New Words and Expressions	(81)
Lesson 10　Material Forming	(82)
Text	(82)
New Words and Expressions	(85)
Notes	(85)
Translation Skills	(86)
Reading Material	(87)
New Words and Expressions	(89)
Lesson 11　Flexible Manufacturing	(90)
Text	(90)
New Words and Expressions	(92)
Notes	(93)
Translation Skills	(93)

  Reading Material ··· (95)
  New Words and Expressions ·· (96)

Lesson 12 Mechatronics ·· (97)
  Text ·· (97)
  New Words and Expressions ·· (98)
  Notes ··· (99)
  Translation Skills ·· (99)
  Reading Material ··· (102)
  New Words and Expressions ·· (103)

Lesson 13 Hydraulic and Pneumatic Systems ·············· (104)
  Text ·· (104)
  New Words and Expressions ·· (108)
  Notes ··· (108)
  Writing Training ··· (109)
  Reading Material ··· (110)
  New Words and Expressions ·· (113)

Lesson 14 Modern Design Theories and Methods ······ (114)
  Text ·· (114)
  New Words and Expressions ·· (118)
  Writing Training ··· (118)
  Reading Material ··· (120)
  New Words and Expressions ·· (121)

Lesson 15 Computer-Integrated Manufacturing ·········· (122)
  Text ·· (122)
  New Words and Expressions ·· (124)
  Notes ··· (124)
  Writing Training ··· (124)
  Reading Material ··· (126)
  New Words and Expressions ·· (128)

Lesson 16 Industrial Robots ······································ (129)
  Text ·· (129)
  New Words and Expressions ·· (131)
  Notes ··· (131)
  Writing Training ··· (132)
  Reading Material ··· (133)
  New Words and Expressions ·· (135)

Lesson 17 Wastewater Treatment Facilities ·············· (136)
  Text ·· (136)
  New Words and Expressions ·· (140)
  Notes ··· (140)

  Writing Training ··· (141)
  Reading Material ·· (142)
  New Words and Expressions ··· (145)
Lesson 18 Remanufacturing Engineering ··· (146)
  Text ·· (146)
  New Words and Expressions ··· (149)
  Notes ··· (149)
  Writing Training ··· (150)
  Reading Material ·· (151)
  New Words and Expressions ··· (153)
Lesson 19 Mining Metallurgy Equipment ·· (154)
  Text ·· (154)
  New Words and Expressions ··· (156)
  Notes ··· (156)
  Writing Training ··· (157)
  Reading Material ·· (158)
  New Words and Expressions ··· (160)
Lesson 20 Sustainable Product Design ··· (161)
  Text ·· (161)
  New Words and Expressions ··· (165)
  Notes ··· (166)
  Writing Training ··· (166)
  Reading Material ·· (167)
  New Words and Expressions ··· (170)
Lesson 21 Automobile Engineering ·· (171)
  Text ·· (171)
  New Words and Expressions ··· (173)
  Notes ··· (173)
  Writing Training ··· (174)
  Reading Material ·· (175)
  New Words and Expressions ··· (177)
Lesson 22 Engineering Equipment ··· (178)
  Text ·· (178)
  New Words and Expressions ··· (182)
  Notes ··· (182)
  Writing Training ··· (183)
  Reading Material ·· (186)
  New Words and Expressions ··· (188)
Lesson 23 Energy Saving Equipment ·· (189)
  Text ·· (189)

New Words and Expressions ················· (191)
　　Notes ······································· (191)
　　Interview Skills ···························· (192)
　　Reading Material ···························· (194)
　　New Words and Expressions ················· (196)
Lesson 24　Product Control and Quality Assurance ················· (197)
　　Text ······································· (197)
　　New Words and Expressions ················· (199)
　　Notes ······································· (200)
　　Interview Skills ···························· (200)
　　Reading Material ···························· (201)
　　New Words and Expressions ················· (202)
Lesson 25　Agricultural Equipment ················· (203)
　　Text ······································· (203)
　　New Words and Expressions ················· (207)
　　Writing Training ···························· (207)
　　Reading Material ···························· (209)
　　New Words and Expressions ················· (210)
Lesson 26　Quality and Environmental Management Systems ················· (211)
　　Text ······································· (211)
　　New Words and Expressions ················· (215)
　　Notes ······································· (215)
　　Writing Training ···························· (216)
　　Reading Materials ··························· (218)
　　New Words and Expressions ················· (221)
附录　参考译文 ································ (222)
　　第 1 课　工程制图 ···························· (222)
　　第 2 课　力学 ······························· (225)
　　第 3 课　工程材料 ···························· (226)
　　第 4 课　机械设计 ···························· (227)
　　第 5 课　机械零件（Ⅱ） ······················· (230)
　　第 6 课　汽车的基础零件 ······················· (233)
　　第 7 课　制造工艺 ···························· (235)
　　第 8 课　公差与互换性 ························· (236)
　　第 9 课　数控技术 ···························· (239)
　　第 10 课　材料成形 ··························· (241)
　　第 11 课　柔性制造 ··························· (243)
　　第 12 课　机电一体化 ························· (245)
　　第 13 课　液压与气压系统 ······················ (246)
　　第 14 课　现代设计理论与方法 ··················· (249)

第 15 课　计算机集成制造 …………………………………………………………… (252)
第 16 课　工业机器人 ………………………………………………………………… (253)
第 17 课　污水处理设备 ……………………………………………………………… (255)
第 18 课　再制造工程 ………………………………………………………………… (257)
第 19 课　矿山冶金设备 ……………………………………………………………… (259)
第 20 课　可持续产品设计 …………………………………………………………… (261)
第 21 课　汽车工程 …………………………………………………………………… (264)
第 22 课　工程装备 …………………………………………………………………… (266)
第 23 课　节能装备 …………………………………………………………………… (269)
第 24 课　生产控制与质量保证 ……………………………………………………… (270)
第 25 课　农业设备 …………………………………………………………………… (272)
第 26 课　质量和环境管理体系 ……………………………………………………… (275)

参考文献 …………………………………………………………………………………… (279)

Lesson 1　Engineering Drawings

Text

A washing machine and a robotic hand in Figure 1.1 are clearly represented for visual purpose by photographs. However, these products could not be manufactured solely from photographs. Detail drawings must be prepared that note the exact shape, sizes and material composition of each component; assembly drawings that show how the total product is put together by fastening each part in proper sequence are also needed.

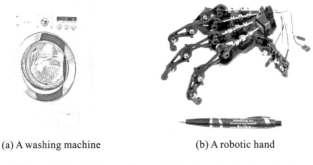

(a) A washing machine　　　　(b) A robotic hand

Figure 1.1　Pictures of some typical mechanical products

Engineering drawing is the means by which engineers design products and create instructions for manufacturing parts. An engineering drawing can be a hand drawing or computer model showing all the dimensions necessary to manufacture a part, as well as assembly notes, a list of required materials, and other pertinent information. Engineering drawings are legal documents, so they must be formal and precise.

In modern manufacturing industry, several types of drawings are acceptable. However, the standard engineering drawing is the multi-view drawing. An engineering drawing usually contains two or three views (front, top, and side). Each view is an orthographic projection of objects (see Figure 1.2). The projection on the frontal plane (x-z) is fixed and the image is called front view. With the projected image, the horizontal plane (x-y) is rotated 90° clockwise on the x axis; the result is the top view of the object. The profile plane (y-z) is rotated 90° clockwise about z axis to obtain the side view.

Line types and conventions　Line types and conventions of engineering drawings are shown in Figure 1.3. Acceptable quality of a drawing is dependent on the density and uniformity of line work (and lettering). Types of lines described herein are merely line conventions, but in every case, each type of line shall be opaque and of uniform width and shall be used on all drawings other than diagrams, such as schematics, etc.

Scales　Drawings shall be made to full scale unless the parts (or assembly) are too large to permit it or so small and complex that drawing to an enlarged scale is essential for clarity. When the main views of large parts are drawn to a reduced scale, the detail views "taken" to clarify detail should be made to full scale whenever possible. When the part has been drawn to an enlarged scale for clarity, it is not necessary to make an actual-size view. The scales preferred for engineering drawings are full size 1 : 1, reduced 1 : 2, 1 : 4, 1 : 10, 1 : 20, and enlarged 2 : 1, 4 : 1, 10 : 1, 20 : 1.

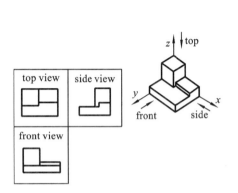

Figure 1.2　Multi-view drawing of an object

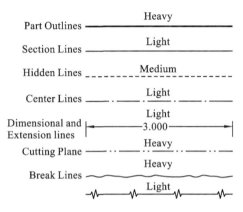

Figure 1.3　Line types and conventions

Sketch drawings　A sketch is a quickly executed freehand drawing that is not intended as a finished work. In general, sketching is a quick way to record an idea for later use. Freehand sketching benefits the entire process in engineering communication: from designers to draftsmen, from workshop supervisors to craftsmen. A sketch may be all that is needed to convey enough of the design that finished engineering drawing can be produced. Sketches may be schematic or instructional and produced to convey ideas between engineering personnel.

Detail drawings　A detail drawing should be a complete and accurate description of a part, with carefully selected views and well-located dimensions of the part. The detail drawing should include all of the necessary information to enable procurement, manufacture and should identify all of the relevant codes and standards. Finished surfaces should be indicated and all necessary shop operations shown. The item weight/mass should also be included for reference. The title should give the material of which the part is to be made and should state the number of the parts that are required for the production of an assembled unit. Detail drawings having only the dimensions and information needed by a particular workman are sometimes made for the different workmen, such as the patternmaker, machinist, or welder.

Assembly drawings　A drawing that shows the parts of a machine or machine unit assembled in their relative working positions is an assembly drawing. A typical assembly drawing may contain the following:

- One or more views, including sections or auxiliaries;
- Enlarged views to show small detail;
- Overall or specific dimensions needed for assembly;
- Notes on manufacturing processes required for assembly;
- Balloons to indicate item numbers;

- Parts list or bill of materials (BOM).

Assembly drawings vary somewhat in character according to their use, such as: design assemblies; working drawing assemblies; genera assemblies, installation assemblies; and check assemblies.

Sectional views Many objects have complicated interior detail which cannot be clearly shown by means of front, top, side, or pictorial views. Sectional views may be necessary in many drawings to bring out and fully dimension the parts. Features of sectional views are cutting-plane symbols (see Figure 1.4), which show where imaginary cutting planes are passed to produce the sections. Usually, one sectional view is sufficient, but several sections may be required for some irregular objects. A full-sectional view or cross-sectional view, showing an object's characteristic shape, usually replaces an exterior front view; however, one of the other principle views, side or top, may be converted to a sectional view if some interior feature thus can be shown to better advantage or if such a view is needed in addition to a sectional front view. Half-sectional view is used when a view is needed showing both the exterior and interior construction of a symmetrical object. A broken-sectional view is used mainly to expose the interior of objects so constructed that less than a half section is required for a satisfactory description.

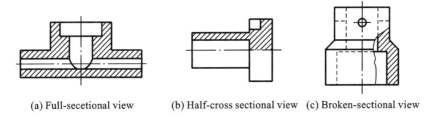

(a) Full-secetional view   (b) Half-cross sectional view   (c) Broken-sectional view

Figure 1.4 Sectional views

Dimensioning Engineering drawings communicate not only geometry (shape and location) of objects but also dimensions and tolerances for those characteristics. Several systems of dimensioning have evolved. The simplest dimensioning system just specifies distances between points (such as an object's length or width, or hole center locations). Since the advent of well-developed interchangeable manufacture, these distances have been accompanied by tolerances of the plus-or-minus or min-and-max-limit types. In drawings of large structures the major unite is the foot, and in drawings of small objects the unit is the inch. In metric dimensioning, the basic unit may be the meter, the centimeter, or the millimeter, depending upon the size of object or structure.

Exploded drawings An exploded drawing is a type of pictorial drawing designed to show several parts in their proper location prior to assembly (see Figure 1.5). Dimensions and relative sizes of items may be shown to indicate mechanical relationship. Although exploded views are not used as working drawings for the machinist, it has an important place in mechanical technology. Exploded views appear extensively in manuals and handbooks that are used for repair and assembly of machines and other mechanisms.

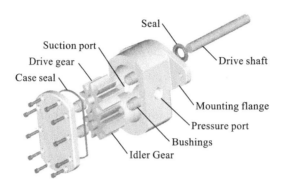

Figure 1.5　Exploded drawing of a gear pump

Schematic drawings　A schematic or diagrammatic drawing show all significant components, parts, or tasks (and their interconnections) of a circuit, device, flow, process, or project by means of symbols. For example, an electrical schematic is a functional schematic which defines the interrelationship of the electrical elements in a circuit, equipment, or system. The symbols describing the electrical elements are stylized, simplified, and standardized to the point of universal acceptance. In a mechanical schematic, the graphical descriptions of elements of a mechanical system are more complex and more intimately interrelated than the symbolism of an electrical system and so the graphical characterizations are not nearly as well standardized or simplified. A process flow schematic is a diagram commonly used in engineering to indicate the general flow of plant processes and equipment. It displays the relationship between major equipment of a plant facility and does not show minor details such as piping details and designations. Schematic diagrams are used extensively in repair manuals to help users understand the interconnections of parts, and to provide graphical instruction to assist in taking apart and rebuilding mechanical assemblies.

Layout drawings　A layout drawing is a graphical statement of the overall form of a component or device, which is usually prepared during the innovative stages of a design. Since it lacks detail and completeness, a layout drawing provides a faithful explanation of the device and its construction only to individuals such as designers and drafters who have been intimately involved in the conceptual stage. In most cases the layout drawing ultimately becomes the primary source of information from which detail drawings and assembly drawings are prepared by other drafters under the guidance of the designer.

New Words and Expressions

1. assembly drawing　装配图
2. balloon　零件序号
3. detail drawing　零件图
4. view　视图
5. full-section view　全剖视图
6. broken-sectional view　局部视图
7. convention　惯例，规范
8. cutting-plane　剖切面
9. sketch drawing　草图
10. interchangeable　可互换的

11. craftsman 工匠，工人
12. tolerance 公差
13. procurement 采购，获得
14. exploded drawing 分解示图
15. patternmaker 模型工，翻铸工
16. machinist 机械师，机工

Notes

(1) An engineering drawing usually contains two or three views (front, top, and side). Each view is an orthographic projection of objects.

工程图样一般含有2~3个视图（主视图、俯视图和侧视图），每个视图均为所绘零件的正投影。

(2) Features of sectional views are cutting-plane symbols, which show where imaginary cutting planes are passed to produce the sections.

剖视图的特征是带有剖面线，剖面线符号表示利用想象的剖切面产生剖视图的位置。

(3) Since the advent of well-developed interchangeable manufacture, these distances have been accompanied by tolerances of the plus-or-minus or min-and-max-limit types.

由于互换性制造技术的出现和不断完善，零件图的尺寸都用带有正、负公差或最大、最小极限的尺寸标注。

(4) A process flow schematic is a diagram commonly used in engineering to indicate the general flow of plant processes and equipment.

工艺流程图常用来说明企业的生产工艺和设备流程。

(5) In most cases the layout drawing ultimately becomes the primary source of information from which detail drawings and assembly drawings are prepared by other drafters under the guidance of the designer.

在大多数情况下，设计人员绘制的总体布局图最终成为其他人员绘制零件图和装配图的第一手资料。

❖ Translation Skills

专业英语的特点与翻译标准

一、专业英语的特点

专业英语属现代英语的科技语域，是科技人员在科技领域交际活动中所用的英语语言。构成语域差别的重要因素包括语言标记、语法、词汇和语音等，不同语域其语言材料不同，如机械、电子、医学和法律等领域均具有特定的语域。专业英语具有以下特点：结构严谨、主题单一、说理明确、大量使用技术术语、图表和符号等。目前，国际上85%的科技文献是用英文发表的。

（一）专业英语词汇特点

大量使用科技术语是专业英语的基本特点，因为专业术语语义单一严谨，为了科学想象、揭示客观事物的发展规律，专业英语必须使用表意确切的专业词汇。专业英语词汇是现代英

语词汇体系的重要组成部分,其形成与发展受现代英语制约,又对现代英语词汇有重要影响。

1. 词汇分类

专业词汇:在科技文献中使用的专业名词和术语,如 mechanical engineering(机械工程)、electronics(电子学)、computer(计算机)、biology(生物学)等。它们用来表达该专业中某一具有确切定义的概念,含义专一。

次专业词汇:这类词汇源于一般英语词汇,但随着科学技术的发展,其意义被扩展,在科技文献中使用既可能有其原来的含义,又可能用作专门的科技术语,如 conductor(指导,导体)、power(能力,功率),work(工作,功),reduction(减少,还原)。

功能词汇:介词、连词等具有一定语法功能的词,如 at,in,on,of,but,and,if 等。

一般词汇:日常生活及任意文体中大量出现的词,如 big,small,come,go,have 等。

2. 词汇特点

1) 词义严谨、规范

比较下列两组同义词:

一般英语	专业英语
to find out	to determine
enough	sufficient
to have	to possess
to use	to employ

2) 前、后缀多

专业英语词汇多数是由拉丁语和希腊语的词根构成的,一般学习者只注意学习如 sun,moon,star,earth,water,stone,life 等这样一些明义词汇,而忽视和它们具有相同意义的"暗"的词根,如 sol,lun,astro,geo,hydr(o),lith,bio 等。利用这些词根,举一反三,能迅速扩大词汇量。随着科学技术的发展,涌现出大量新词。专业英语使用前、后缀构词比一般英语多,如 auto-mobile(汽车),semi-conductor(半导体),micro-economics(微观经济学)等。

3) 外来语、缩略语多

据统计,科技英语文体中有 2 万~3 万的缩略语,另外,希腊语、拉丁语在科技英语中的应用也很多。

外来语如:etc.[etcetera,拉丁语]＝and the rest,and so on (等等);et al.[拉丁语]＝and the other people(等人)。

缩略语如:CAD[Computer Aided Design](计算机辅助设计)。

3. 常用构词法

1) 派生词汇

派生词汇是由词缀与词根结合构成的词汇,主要源于希腊语和拉丁语,例如:

Hydro-　相当于 water,如 hydrology(水文学),dehydration(脱水),hydrodynamics(流体动力学);

-ject-　相当于 to throw(投,掷),如 eject(喷出),inject(注入),reject(拒绝)。

2) 合成词汇

A:复合名词

名词＋名词:ballbearing (球轴承);crankshaft (曲轴);flowchart(流程图)

专有名词＋名词:Laplas transformation (拉普拉斯变换);Greenwich time (格林威治时

间）

名词所有格+名词：Archimedes' principle（阿基米德原理）；Hook's law（胡克定律）

形容词+名词：blueprint（蓝图）；deadweight（船舶载重量）

大写英文字母+名词：U-tube（U形管）；V-belt（V带）；I-beam（工字梁）；T-slot（T形槽）

希腊字母（英文音）+名词：Gamma ray（γ射线）；Alpha-activity（放射性）

B：复合形容词

形容词+名词：*quick-action* valve（速动阀）；*low-carbon* steel（低碳钢）；*upper-limit* speed（上限速度）

名词+形容词：*temperature-sensitive* element（热敏元件）；*energy-efficient* engine（节能型发动机）

名词+名词：*pressure-control* valve（压力控制阀）；*heat-dissipation* problem（热量耗散问题）

名词+分词：*sound-absorbing* material（吸音材料）；*motor-driven* equipment（电力拖动设备）

形容词+分词：*far-ranging* missile（远程导弹）；*cold-drawn* bar（冷拔棒料）

副词+分词：*newly-developed* device（新开发的装置）；*ever-increasing* demand（不断增长的需求）

数词+单位名词+形容词：*36-inch-long* pipe（36英寸长的管道）

数词+名词：*eight-cylinder* engine（八缸发动机）；*double-row* ballbearing（双列球轴承）

数词+形容词：*three-dimensional* image（三维图像）

3）缩略词汇

A：截短

即将词的某个音节省略或简化，如 lab=laboratory（实验室），Jan.=January（一月）。

B：首字母缩略

即将一个词组的每个词的第一个字母组合，如 UFO=unidentified flying object（不明飞行物），laser=light amplification by stimulated emission of radiation（激光）。

(二) 专业英语的语法和句法特点

1. 复杂长句多

科技文章要求叙述准确、推理严谨、结构紧凑，因此会往往出现很多长句。复杂长句一般分为两类：一类是带有较多的定语和状语的长句，另一类是包含多个从句或分句的复合长句。

[例1] An industrial robot has described by the International Organization for Standardization (ISO) as "a machine formed by mechanism including several degrees of freedom, often having the appearance of one or several arms ending in a wrist capable of holding a tool, a work piece, or an inspection device".

译文：国际标准组织（ISO）对工业机器人的定义是："一种由包含多个自由度的机构组成的机器，这种机器通常含有一个或几个手臂，这些手臂末端有腕关节，能够抓取工具、工件或者监控设备。"

[例2] The process of automating a manufacturing process may take many forms ranging from a rudimentary level, perhaps involving a simple mechanical device that operations part of a machine, to complex computer driven feedback systems that have considerable decision making capacity over the process that they controlling.

译文：制造过程的自动化可以采用多种形式，从可能包括作为机械操作部分的简单设备的基础装置，到复杂的计算机驱动的反馈系统（这些系统在它们控制的加工过程中具有相当大的决策能力）均可采用。

2. 非谓语动词多

科技文体要求文笔简练、结构紧凑，为此大量使用非限定动词代替定语从句、状语从句等。

［例1］ *Finding* the resultant of several forces is called composition of forces.

译文：确定几个力的合力叫做力的合成。

［例2］ Materials *to be used for structural purposes* are chosen so as to behave elastically in the environmental conditions.

译文：结构材料的选择应使其在外界条件中保持弹性。

3. 名词化结构多

科技文体要求表达客观、内容确切、信息量大，强调存在的事实，大量使用名词化结构是专业英语的特点之一。

［例］ Television is *the transmission and reception of images of moving objects by radio waves*.

译文：电视通过无线电波发射和接收活动物体的图像。

4. 后置定语多

使用动词的非谓语形式和形容词短语作后置定语也是科技文体的特点之一。例如：

［例1］ A machine is composed of rigid bodies *having definite motions and capable of performing useful work*.

译文：一部机器一般由刚性机构组成，这些机构具有特定的运动形式，能够完成有效的工作。

［例2］ The heat *produced* is equal to the electrical energy *wasted*.

译文：产生的热量等于浪费的电能。

5. 被动语态多

科技文体侧重叙事推理，强调客观事物，广泛采用被动语态。例如：

Individually manufactured parts and components are assembled automatically into subassemblies and assemblies to form a product.

译文：制造出的单个零件自动进行局部装配和总体装配，从而形成产品。

二、专业英语的翻译标准和翻译过程

专业英语文体的特点是由科技文献的内容所决定的，专业英语的翻译也有别于其他英语文体的翻译。

对专业英语的翻译，要求译文达到忠实、通顺和简练的标准。"忠实"，即要求译文忠实于原文，准确、完整和科学地表达原文的内容，包括思想、原意与风格，译者不能任意对原文内容加以歪曲、增删、遗漏或篡改，例如，assembly 在机械工程英语中为"装配、组装"，而非"集合、集会"之意；"通顺"要求译文语言必须通顺，符合汉语规范，用词造句符合汉语语言习惯，以求通顺易懂，不应有文理不通、逐词死译和生硬晦涩等现象（见下文中例1和例2）；"简练"要求译文语言表达简洁、优美。

[例 1] We must insert a new fuse every time it has *functioned*.
原译:每当熔丝起作用之后,必须插进一个新的。
改译:每当熔丝烧断之后,必须插进一个新的。

[例 2] The shortest distance between raw material and a finished part is casting.
原译:原材料和成品的最短距离是铸造。
改译:将原材料转化成成品的捷径是铸造。

为了提高专业英语的翻译质量,使译文准确、通顺、简练,译者需要善于运用专业英语的翻译技巧和方法,如词义引申、增减、词类转换等(将在后续课程中讲解),同时要求译者具备高的英语水平、汉语水平和专业水平,在翻译过程中要求正确理解原文内容和语境、准确表达原文思想和对译文进行认真校对。

[例 3] Such materials are *characterized* by good insulation and high resistance to wear.
译文:这些材料的特点是绝缘性好、耐磨性强。(动词转化为名词)

❖ Reading Material

Computer-Aided Drafting

The process of preparing engineering drawings on a computer is known as Computer-Aided Drafting (CAD), and it is the most significant development to occur in this field. Manual drawing of a project is often a bottleneck because it takes so much time. In the traditional hand drawing process, drafters spend approximately two-thirds of their time "laying lead". CAD technology was first introduced in the mid-1960 as a tool for the production of drawings without the use of traditional drafting tools. The drawings were created and displayed by manipulating graphic elements on the computer screen instead of drawing them by hand. Today, the drafting tasks have largely been automated and accelerated through the use of CAD system.

CAD can relieve you from many tedious chores such as redrawing. Once you have made a drawing you can store it on a disk. You may then call it up at any time and change it quickly and easily. In manual drawing, you must have the skill to draw lines and letters and use equipment such as drafting tables and machines, and drawing aids such as compasses, protractors, triangles, parallel edges, scales, and templates. In recent years, many educational institutions include CAD as part of their academic curriculum. As a result, CAD knowledge has become very important to all professionals involved in the field of design and drafting.

A CAD computer contains a drafting program that is a set of detailed instructions for the computer. When you bring up the program, the screen displays each function or instruction you must follow to make a drawing. The CAD programs available to you contain all of the symbols used in mechanical, electrical, or architectural drawing. You will use the keyboard and/or mouse to call up the drafting symbols you need. Examples are characters, grid patterns, and types of lines. When you get the symbols you want on the screen, you will order the computer to size, rotate, enlarge, or reduce them, and position them on the screen to produce the image you want. You probably will then order the computer to print the final product and store it for later use. The computer also serves as a filing system for any drawing symbols or completed drawings stored in its memory or on disks. You can call up this information any time and copy it or revise it to produce a different symbol or drawing.

There are two types of CAD systems used for the production of technical drawings—two dimensions (2D) and three dimensions (3D). 2D CAD systems such as AutoCAD or MicroStation replace the paper drawing discipline. The lines, circles, arcs and curves are created within the software. A 2D CAD system is merely an electronic drawing board. Its greatest strength over direct to paper technical drawing is in the making of revisions. Whereas in a conventional hand drawn technical drawing, if a mistake is found, or a modification is required, a new drawing must be made from scratch. The 2D CAD system allows a copy of the original to be modified, saving considerable time. 2D CAD systems can be used to create plans for large projects such as buildings and aircraft but provide no way to check the various components will fit together. 3D CAD systems such as Autodesk Inventor or SolidWorks first produce the geometry of the part; the technical drawing comes from user defined views of the part. Any orthographic, projected and section views can be created by the software. A 3D CAD system allows individual parts to be assembled together to represent the final product(see Figure 1.7). Buildings, aircraft, ships and cars are modeled, assembled and checked in 3D before technical drawings are released for manufacture. Both 2D and 3D CAD systems can be used to produce technical drawings for any discipline. The various disciplines: electrical, electronic, pneumatic, hydraulic, etc., have industry recognized symbols to represent common components.

Figure 1.7　Samples of three-dimensional drafting

Today, engineering design capability was added to many of the CAD programs and thus the name was changed to Computer-Aided Design and Drafting (CADD). CADD is the process where a drafter/designer/engineer creates drawings or models that define a given product before it is ready to be manufactured. The drafter is the key link in the design engineering process and manufacturing steps, and must possess a working knowledge of design principles, material properties, and manufacturing processes. CADD technology has become the preferred method for the preparation, distribution, storage, and maintenance of engineering drawings. With CADD systems, graphic two-dimensional or three-dimensional digital data are placed on various drawing layers that can be selectively displayed and edited.

New Words and Expressions

1. manipulate   熟练操作,操纵
2. Computer-Aided Drafting   计算机辅助绘图
3. protractor   量角器,分度规
4. parallel edge   直尺
5. scale   比例尺,刻度
6. Computer-Aided Design and Drafting   计算机辅助设计及绘图
7. template   模板,样板
8. dimension   维数,尺寸

Lesson 2　Mechanics

Text

Mechanics is, in the most general sense, the study of forces and their effect upon matter. Typically, engineering mechanics is used to analyze and predict the acceleration and deformation (both elastic and plastic) of objects under known forces (also called loads) or stresses. Subdisciplines of mechanics include statics, dynamics, mechanics of materials, kinematics, continuum mechanics.

Mechanical engineers typically use mechanics in the design or analysis phases of engineering. If the engineering project is the design of a vehicle, statics might be employed to design the frame of the vehicle, in order to evaluate where the stresses will be most intense. Dynamics might be used when designing the car's engine, to evaluate the forces in the pistons and cams as the engine cycles. Mechanics of materials might be used to choose appropriate materials for the frame and engine. Fluid mechanics might be used to design a ventilation system for the vehicle, or to design the intake system for the engine.

Statics is a branch of mechanics concerned with the analysis of loads (force, torque/moment) on physical systems in static equilibrium, that is, in a state where the relative positions of subsystems do not vary over time, or where components and structures are at a constant velocity. When in static equilibrium, the system is either at rest, or its center of mass moves at constant velocity.

By Newton's first law, this situation implies that the net force and net torque (also known as moment of force) on every part of the system is zero. From this constraint, such quantities as stress or pressure can be derived. The net forces equaling zero is known as the first condition for equilibrium, and the net torque equaling zero is known as the second condition for equilibrium.

In the field of physics, the study of the causes of motion and changes in motion is dynamics. In other words it is the study of forces and why objects are in motion. Dynamics includes the study of the effect of torques on motion. These are in contrast to kinematics, the branch of classical mechanics that describes the motion of objects without consideration of the causes leading to the motion.

Generally speaking, researchers involved in dynamics study how a physical system might develop or alter over time and study the causes of those changes. In addition, Isaac Newton established the physical laws which govern dynamics in physics. By studying his system of mechanics, dynamics can be understood. In particular dynamics is mostly related to Newton's second law of motion. However, all three laws of motion should be taken into consideration, because these are interrelated in any given observation or experiment.

Strength of materials is a subject in materials science dealing with the ability of objects to withstand stresses without failure. The study of strength of materials often refers to various methods of calculating stresses in structural members, such as beams, columns and shafts. The methods employed to predict the response of a structure under loading and its susceptibility to various failure modes may take into account various properties of the materials other than material yield strength and ultimate strength; for example, failure by buckling is dependent on material stiffness and thus Young's modulus.

Fluid mechanics is the study of fluids and the forces on them. (Fluids include liquids, gases, and plasmas.) Fluid mechanics can be divided into fluid statics, the study of fluids at rest; fluid kinematics, the study of fluids in motion; and fluid dynamics, the study of the effect of forces on fluid motion. It is a branch of continuum mechanics, a subject which models matter without using the information that it is made out of atoms, that is, it models matter from a macroscopic viewpoint rather than from a microscopic viewpoint. Fluid mechanics, especially fluid dynamics, is an active field of research with many unsolved or partly solved problems. Fluid mechanics can be mathematically complex. Sometimes it can best be solved by numerical methods, typically using computers. A modern discipline, called computational fluid dynamics (CFD), is devoted to this approach to solving fluid mechanics problems.

The study of **kinematics** is often referred to as the geometry of motion. To describe motion, kinematics studies the trajectories of points, lines and other geometric objects and their differential properties such as velocity and acceleration. Kinematics is used in astrophysics to describe the motion of celestial bodies and systems, and in mechanical engineering, robotics and biomechanics to describe the motion of systems composed of joined parts (multi-link systems) such as an engine, a robotic arm or the skeleton of the human body.

For a planer motion of a rigid body shown in Figure 2.1, suppose that the motion of point A in the rigid body is shown, as is the angular rotation of the body ω. Point B is an arbtrarily chosen point whose motion we seek. As shown in Figure 2.1 the vectors $\boldsymbol{r}_{AO}$ and $\boldsymbol{r}_{BO}$ denote the positions of points A and B respectively. Note that ω is independent of the choice of points A and B, because the rotation is the same for all lines in a rigid body. The relationship between the absolute motions of points A and B is readily formed by referring to the position vectors depicted in Figure 2.1, i.e.,

$$\boldsymbol{r}_{BO} = \boldsymbol{r}_{AO} + \boldsymbol{r}_{BA} \qquad (1)$$

$$\boldsymbol{v}_B = \boldsymbol{v}_A + \boldsymbol{v}_{BA} = \boldsymbol{v}_A + \omega\, \boldsymbol{r}_{BA} \qquad (2)$$

$$a_B = a_A + a_{BA} = a_A + \alpha\, r_{BA} - \omega^2 r_{BA} \qquad (3)$$

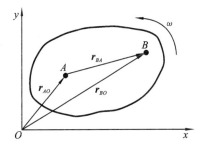

Figure 2.1 The position of vectors

In the above equations (2) and (3), the velocity and acceleration of point A are denoted by v_A and a_A, respectively. The symbols v_B and a_B are similarly defined for point B. The vector ω is the angular velocity of the body, and α is the angular acceleration. Both vectors are perpendicular to the plane of motion, so they are parallel to the axis for the rotational portion of the motion.

Continuum mechanics is a branch of mechanics that deals with the analysis of the kinematics and the mechanical behavior of materials modeled as a continuous mass rather than as discrete particles. The French mathematician Augustin Louis Cauchy was the first to formulate such models in the 19th century, but research in the area continues today.

New Words and Expressions

1. mechanics   力学
2. acceleration   加速度
3. deformation   变形
4. stress   应力
5. subdiscipline   分支学科
6. statics   静力学
7. dynamics   动力学
8. mechanics of materials   材料力学
9. fluid mechanics   流体力学
10. kinematics   运动学
11. continuum mechanics   连续介质力学
12. static equilibrium   静力平衡
13. Newton's first law   牛顿第一定律
14. susceptibility   敏感性
15. Newton's second law of motion
    牛顿运动第二定律
16. yield strength   屈服强度
17. ultimate strength   极限强度
18. failure by buckling   屈曲破坏
19. stiffness   刚度
20. Young's modulus   杨氏模量
21. macroscopic   宏观
22. microscopic   微观
23. computational fluid dynamics (CFD)
    计算流体力学
24. trajectory   轨道
25. astrophysics   天体物理学
26. celestial   天空的
27. robotics   机器人学
28. biomechanics   生物力学
29. rigid body   刚体

Notes

(1) Mechanics is, in the most general sense, the study of forces and their effect upon matter.
就最一般的意义而言，力学研究的是物体的受力及力对物体的影响。

(2) Statics is the branch of mechanics concerned with the analysis of loads (force, torque/moment) on physical systems in static equilibrium, that is, in a state where the relative positions of subsystems do not vary over time, or where components and structures are at a constant velocity.

静力学是力学的一个分支，它主要涉及物理系统在受到载荷（力、扭矩/力矩）处于静力平衡时的分析。静力平衡状态即子系统的相对位置不随时间变化，或其中的成分和结构处在一个恒定的速度下。

(3) It is a branch of continuum mechanics, a subject which models matter without using the information that it is made out of atoms, that is, it models matter from a macroscopic viewpoint rather than from a microscopic viewpoint.

流体力学是连续介质力学的一门分支学科，这个学科并不根据原子提供的信息来考虑物质，也就是说，是从宏观的角度而不是从微观角度来研究物质的。

(4) Continuum mechanics is a branch of mechanics that deals with the analysis of the kinematics and the mechanical behavior of materials modeled as a continuous mass rather than as discrete particles. The French mathematician Augustin Louis Cauchy was the first to formulate such models in the 19th century, but research in the area continues today.

连续介质力学是力学的一个分支，它涉及模型是连续体而不是离散颗粒的材料的力学性能和运动学的分析。在19世纪，法国数学家奥古斯丁·路易斯·柯西第一个用公式表示出了这些模型，但在这一领域的研究一直持续到今天。

❖ Translation Skills

专业英语词汇的构成

一、派生词

(一) 后缀

后缀是在单词后部加上构词结构，形成新的单词。例如：

-scope(探测仪器)，如 microscope(显微镜)，telescope(望远镜)，spectroscope(分光镜)；

-meter(计量仪器)，如 barometer(气压表)，telemeter(测距仪)，spectrometer(分光仪)；

-graph(记录仪器)，如 tomograph(X射线体层照相)，telegraph(电报)，spectrograph(分光摄像仪)；

-able(可能的)，如 enable(允许、使能)，disable(禁止、不能)，programmable(可编程的)；portable(便携的)，scalable(可缩放的)；

-ware(件/部件)，如 hardware(硬件)，software(软件)，firmware(固件)，groupware(组件)，freeware(免费软件)；

-ity(性质)，如 reliability(可靠性)，availability(可用性)，accountability(可核查性)，integrity(完整性)，confidentiality(保密性)。

(二) 前缀

采用前缀构成的单词在计算机专业英语中占了很大比例，通过下面的实例可以了解这些由常用的前缀构成的单词。

multi-(多)，如 multiaxial(多轴的)；

hyper-(超级)，如 hypercube 超立方；

super-(超级)，如 superhighway 超级公路；

inter-(相互、在……间)，如 interface(接口、界面)，interlace(隔行扫描)；

micro-(微型)，如 microprocessor(微处理器)，microkernel(微内核)；

tele-(远程的)，如 telephone(电话)，teletext(图文电视)。

单词前缀还有很多，其构成可以同义而不同源(如拉丁语、希腊语)，可以互换，例如：

multi-，poly-相当于 many，如 multimedia(多媒体)，polytechnic(各种工艺的)；

uni-，mono-相当于 single，如 unicode(统一的字符编码标准)，monochrome(单色)；

bi-，di-相当于 twice，如 binomial(二项式)，dibit(双位)；

equi-，iso-相当于 equal，如 equality(等同性)，isochromatic(等色的)；

simil-，homo-相当于 same，如 similarity(类似)，homogeneous(同类的)；

semi-，hemi-相当于 half，如 semiconductor(半导体)，hemicycle(半圆形)。

二、复合词

复合词是科技英语中另一大类词汇，其组成面广，通常分为复合名词、复合形容词、复合动词等。复合词通常以连字符"-"连接单词构成，或者采用短语构成。如：

-based(基于，以……为基础)，如 rate-based(基于速率的)，credit-based(基于信誉的)；

-oriented(面向……的)，如 object-oriented(面向对象的)，market-oriented(市场导向的)；

info-(信息，与信息有关的)，如 info-war(信息战)，info-tree(信息树)；

centric(以……为中心的)，如 client-centric(以客户为中心的)，user-centric(以用户为中心的)；

-free(自由的，无关的)，如 lead-free(无线的)，jumper-free(无跳线的)。

此外，以名词＋动词-ing 构成的复合形容词形成了一种典型的替换关系，即可以根据需要在结构中代入同一词类而构成新词，它们多为动宾关系。如：

man-carrying aircraft(载人飞船)，earth-moving machine(推土机)，time-consuming operation(耗时操作)，ocean-going freighter(远洋货舱)。

然而，必须注意，复合词不可以随意构造，否则可能形成一种非正常的英语句子结构。虽然上述例子给出了多个连接单词组成的复合词，但不提倡这种冗长的复合方式。对于由多个单词加连字符组成的词组，要注意单词的顺序和主要针对对象。此外还应当注意，有时加连字符的复合词与不加连字符的词汇词意是不同的，必须通过文章的上下文推断，如 force-feed 表示强迫接受(vt.)，而 force feed 则为加压润滑。

随着词汇的专用化，有些复合词中间的连字符被省略掉，形成了一个单词，例如：

videotape(录像带)   fanin(扇入)   fanout(扇出)

online(在线)   onboard(板载)   login(登录)

logout(撤销)   pushup(拉高)   popup(弹出)

三、混成词

混成词不论在公共英语中还是科技英语中都大量存在，也有人将其称为缩合词(与缩略词区别)、融会词。混成词多是名词，也有的被作为动词使用，对这类词汇可以通过其构词规律和词素进行理解。所谓混成，是将一个单词的前部与另一个单词的后部拼接，或者将两个单词的前部拼接，或者将一个单词前部与另一个词拼接构成新的词汇，例如：

brunch (breakfast+lunch) 早中饭
smog (smoke+fog) 烟雾
codec (coder+decoder) 编码译码器
compuser (computer+user) 计算机用户
transeiver (transmitter+receiver) 收发机
syscall (system+call) 系统调用
mechatronics (mechanical+electronic) 机械电子学
calputer (calculator+computer) 计算器式电脑

四、缩略词

缩略词是将较长的英语单词取其首部或者主干构成与原词同义的短单词,或者将组成词汇短语的各个单词的首字母拼接为一个大写字母的字符串。随着科技发展,缩略词在文章索引、前言、摘要、文摘、电报、说明书、商标等科技文章中被频繁采用。缩略词的构词方法有以下两种。

(一) 压缩和省略

将某些太长、难拼、难记、使用频繁的单词压缩成一个短小的单词,或取其头部,也可取其关键音节。例如:

flu=influenza(流感)　　lab=laboratory(实验室)
math=mathematics(数学)　　iff=if only if(当且仅当)
rhino=rhinoceros(犀牛)　　ad=advertisement(广告)

(二) 缩写

将某些词组和单词集合中每个实义单词的第一或者首部几个字母重新组合,组成为一个新的词汇,作为专用词汇使用。在应用中它形成了三种类型。

(1) 以小写字母出现,并作为常规单词,例如:

flops (floating-point operation per second) 每秒浮点运算次数
spool (simultaneous peripheral operation on line) 假脱机

(2) 以大写字母出现,具有主体发音音节,例如:

BASIC (beginners' all-purpose symbolic instruction code) 初学者通用符号指令代码
FORTRAN (formula translation) 公式翻译(语言)
COBOL (common business oriented language) 面向商务的通用语言

(3) 以大写字母出现,没有读音音节,仅为字母头缩写,例如:

RISC (reduced instruction set computer) 精简指令集计算机
CISC (complex instruction set computer) 复杂指令集计算机
ADE (application development environment) 应用开发环境
PCB (process control block) 进程控制块
CGA (color graphics adapter) 彩色图形适配器

五、借用词

借用词一般来自厂商名、商标名、产品代号名、发明者名、地名等,它通过将普通公共英语词汇演变成专业词义而实现,有的则是原来已经有的词汇,只是被赋予了新的含义。

例如：

woofer(低音喇叭)　　tweeter(高音喇叭)　　flag(标志、状态)
cache(高速缓存)　　semaphore(信号量)　　firewall(防火墙)
mailbomb(邮件炸弹)　scratchpad(高速缓存)　fitfall(专用程序入口)

现代科技英语借用了大量的公共英语词汇、日常生活中的常用词汇，读者必须在努力扩大自己专业词汇的同时，丰富自己的生活词汇，并在阅读和翻译时正确采用词义。

❖ Reading Material

Continuum Mechanics

Modeling an object as a continuum assumes that the substance of the object completely fills the space it occupies. Modeling objects in this way ignores the fact that matter is made of atoms, and so is not continuous; however, on length scales much greater than that of inter-atomic distances, such models are highly accurate. Fundamental physical laws such as the conservation of mass, the conservation of momentum, and the conservation of energy may be applied to such models to derive differential equations describing the behavior of such objects, and some information about the particular material studied is added through a constitutive relation.

Continuum mechanics deals with physical properties of solids and fluids which are independent of any particular coordinate system in which they are observed. These physical properties are then represented by tensors, which are mathematical objects that have the required property of being independent of coordinate system. These tensors can be expressed in coordinate systems for computational convenience.

Materials, such as solids, liquids and gases, are composed of molecules separated by empty space. On a macroscopic scale, materials have cracks and discontinuities. However, certain physical phenomena can be modeled assuming the materials exist as a continuum, meaning the matter in the body is continuously distributed and fills the entire region of space it occupies. A continuum is a body that can be continually sub-divided into infinitesimal elements with properties being those of the bulk material.

The validity of the continuum assumption may be verified by a theoretical analysis, in which either some clear periodicity is identified or statistical homogeneity and ergodicity of the microstructure exists. More specifically, the continuum hypothesis/assumption hinges on the concepts of a representative volume element (RVE) (sometimes called "representative elementary volume") and separation of scales based on the Hill-Mandel condition. This condition provides a link between an experimentalist's and a theoretician's viewpoint on constitutive equations (linear and nonlinear elastic/inelastic or coupled fields) as well as a way of spatial and statistical averaging of the microstructure.

Major areas of continuum mechanics

Continuum mechanics: The study of the physics of continuous materials	Solid mechanics: The study of the physics of continuous materials with a defined rest shape	Elasticity: Describes materials that return to their rest shape after an applied stress	
		Plasticity: Describes materials that permanently deform after a sufficient applied stress	Rheology: The study of materials with both solid and fluid characteristics
	Fluid mechanics: The study of the physics of continuous materials which take the shape of their container	Non-Newtonian fluids: a type of fluid whose flow properties differ in any way from those of Newtonian fluids	
		Newtonian fluids: a fluid whose stress versus strain rate curve is linear and passes through the origin	

When the separation of scales does not hold, or when one wants to establish a continuum of a finer resolution than that of the RVE size, one employs a statistical volume element (SVE), which, in turn, leads to random continuum fields. The latter then provides a micromechanics basis for stochastic finite elements (SFE). The levels of SVE and RVE link continuum mechanics to statistical mechanics. The RVE may be assessed only in a limited way via experimental testing: when the constitutive response becomes spatially homogeneous.

Specifically for fluids, the Knudsen number is used to assess to what extent the approximation of continuity can be made.

New Words and Expressions

1. separation of scales 尺度分离
2. conservation of mass 质量守恒定律
3. experimentalist 实验主义者
4. conservation of momentum 动量守恒定律
5. theoretician 理论家
6. conservation of energy 能量守恒定律
7. statistical volume element 统计体积元素
8. differential equation 微分方程式
9. micromechanics 微观力学
10. constitutive relation 本构关系
11. Knudsen number 努森数
12. tensor 张量
13. solid mechanics 固体力学
14. coordinate system 坐标系
15. fluid mechanics 流体力学
16. crack 裂纹
17. elasticity 弹性力学
18. discontinuity 不连续
19. plasticity 塑性力学
20. infinitesimal 无穷小
21. rheology 流变学
22. homogeneity 均一性
23. non-Newtonian fluid 非牛顿流体
24. ergodicity 遍历性
25. Newtonian fluid 牛顿流体
26. representative volume element 典型体积元素

Lesson 3 Engineering Materials

Text

The history of civilization evolved from the Stone Age to the Bronze Age, the Iron Age, the Steel Age, and to the Space Age (contemporaneous with the Electronic Age). Each age is marked by the advent of certain materials. Today, for industrial purposes, materials are divided into engineering materials or nonengineering materials. Engineering materials are those used in manufacture and become parts of products. Nonengineering materials are the chemicals, fuels, lubricants, and other materials used in the manufacturing process, which do not become part of the product. Engineering materials constitute the foundation of technology whether the technology pertains to structural, electronic, thermal, electrochemical, environmental, biomedical, or other applications. The principal classes of engineering materials include metals, polymers, ceramics, composite materials, etc. (see Figure 3.1).

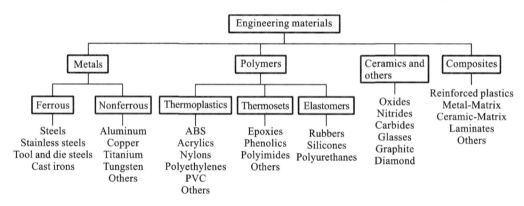

Figure 3.1 An outline of engineering materials

Metals (including alloys)

Materials in this group are composed of one or more metallic elements (e. g. , Fe, Al, Cu, Ti, Au, and Ni), and often also nonmetallic elements (e. g. , C, N, and O) in relatively small amounts. Atoms in metals and their alloys are arranged in a very orderly manner, and in comparison to the ceramics and polymers, are relatively dense. With regard to mechanical characteristics, these materials are relatively stiff and strong, yet are ductile (i. e. , capable of large amounts of deformation without fracture), and are resistant to fracture, which accounts for their widespread use in structural applications. Metallic materials have large numbers of delocalized electrons; that is, these electrons are not bound to particular atoms. Many properties of metals are directly attributable to these electrons. For example, metals are extremely good conductors of electricity and heat, and are not transparent to visible light; a polished metal surface has a lustrous appearance. In addition, some of the metals (i. e. , Fe, Co, and Ni) have desirable magnetic properties.

While metals comprise about three-fourths of the elements that we use, few find service in their pure form. There are several reasons for not using pure metals. Pure metals may be too weak or too soft, or they may be too costly because of their scarcity, but the key factor normally is that desired property sought in engineering requires a blending of metals and other elements. Thus, the combination forms (alloys) find greatest use.

Metal alloys, by virtue of composition, are often grouped into two classes—ferrous and nonferrous. Ferrous alloys, those in which iron is the principal constituent, include steels and cast irons. They are especially important as engineering construction materials making up the largest proportion both by quantity and commercial value. Their widespread use is accounted for by three factors: (1) iron-containing compounds exist in abundant quantities within the earth's crust; (2) metallic iron and steel alloys may be produced using relatively economical extraction, refining, alloying, and fabrication techniques; and (3) ferrous alloys are extremely versatile, in that they may be tailored to have a wide range of mechanical and physical properties.

Steel is our most widely used alloy. Iron alloyed with various proportions of carbon gives low, mid and high carbon steels. An iron carbon alloy is considered steel only if the carbon level is between 0.01% and 2.00%. For the steels, the hardness and tensile strength of the steel is related to the amount of carbon present, with increasing carbon levels also leading to lower ductility and toughness. However, heat treatment processes such as quenching and tempering can significantly change these properties. The most common types of steels are plain low-carbon, high-strength low-alloy, medium-carbon, tool, and stainless. Stainless steel is defined as a regular steel alloy with greater than 10% by weight alloying content of Chromium. Nickel and Molybdenum are typically also found in stainless steels. A wide range of mechanical properties combined with excellent resistance to corrosion make stainless steels very versatile in their applicability. Cast iron is defined as an iron-carbon alloy with more than 2.00% but less than 6.67% carbon.

All alloys that are not iron-based are nonferrous. The significant nonferrous alloys include those of copper, aluminum, titanium, and magnesium. Copper and copper-based alloys, possessing a desirable combination of physical properties, have been utilized in quite a variety of applications such as piping, utensils, thermal conduction, electrical conduction, etc. The alloys of aluminum, titanium and magnesium are known and valued for their high strength-to-weight ratios and, in the case of magnesium, their ability to provide electromagnetic shielding. These materials are ideal for situations where high strength-to-weight ratios are more important than bulk cost, such as in the aerospace industry and certain automotive engineering applications.

New Words and Expressions

1. polymer　聚合物
2. ceramics　陶瓷
3. stiff　硬的,刚性的
4. fracture　断裂,折断
5. transparent　透明的,显然的
6. lustrous　有光泽的
7. delocalized　不受位置限制的
8. ferrous　铁的,含铁的
9. nonferrous　不含铁的,非铁的
10. tailored　定制的,特制的
11. hardness　硬度
12. tensile strength　抗拉强度
13. toughness　韧性
14. quenching　淬火
15. tempering　回火
16. stainless　不锈的
17. shield　防护,屏蔽,遮挡

Notes

(1) With regard to mechanical characteristics, these materials are relatively stiff and strong, yet are ductile (i.e., capable of large amounts of deformation without fracture), and are resistant to fracture, which accounts for their widespread use in structural applications.

就力学性能而言,这些材料比较硬而且坚固,但是可延伸(即能够产生大的变形量而不断裂),可以抵抗断裂,这些是它们被广泛用在结构性应用中的原因。

(2) Pure metals may be too weak or too soft, or they may be too costly because of their scarcity, but the key factor normally is that desired property sought in engineering requires a blending of metals and other elements. Thus, the combination forms (alloys) find greatest use.

纯金属可能强度太差或者太软,也可能因为稀少而太昂贵,但关键的因素通常是,为获得在工程中所追求的理想性能,需要将金属和其他元素混合起来。因此,这种结合的形式(合金)就得到了很广泛的应用。

❖ Translation Skills

词性的转换

英语和汉语分属两个不同的语系,在语言的表达方式和内容上都存在着很大的差异。英语和汉语最大区别之一是英语讲究词性。如果在翻译过程中,把英语句子中的词性、句式照搬过来,则汉语译文会变得生硬蹩脚,使读者不易理解。因此,在翻译中,经常采用词性转换的翻译技巧,也就是把英语中的某种词性,转换成汉语的另外一种词性来表达,以使译文符合汉语的表达习惯。在必要的时候,词性转换几乎可以在所有词性间进行,这种转换会使译文更加灵活、贴切,不拘泥于原文形式,却能更准确地传达出原文内涵。常见的词性转换形式有如下四种。

一、转译成动词

与英语相比较而言,汉语中动词用得较多,在英语句子中往往只有一个谓语动词。英语还可以通过谓语动词以外各种词性的词来体现动词意义。所以,英语中不少词类在翻译时可以转译成动词。

(一) 名词转译成动词

英语中有大量由动词派生的名词和具有动作意义的名词以及其他名词,往往可将其转译成动词。

[例1] In industry field, there is a lot of *emphasis* on saving energy.

译文:在工业领域中,人们非常重视节约能源。

[例2] One of the main streams of energy saving technologies is the *reduction* of power consumption while idling operation.

译文:能量节约技术的主流之一是减少在空转中的能量消耗。

[例3] A careful *study* of the original paper will make you understand better.

译文:仔细研究原文,你会更好地理解它。

[例4] Those small factories are also lavish *consumers* and *wasters* of raw materials.

译文:那些小工厂同样在极大地消耗和浪费原材料。

(二) 形容词转译成动词

英语中形容词常与系动词搭配构成系表结构,系表结构作谓语,从而使这些形容词具有了动作意味。这些形容词常常是表示知觉、情感、欲望等与心理状态相关的形容词。

[例1] Engineers are *concerned* about the safety of these machines.

译文:工程师关注这些机器的安全性。

[例2] We are not *content* with our present achievements.

译文:我们不满足于我们现有的成就。

(三) 副词转译成动词

[例1] After careful investigation they found the design *behind*.

译文:在仔细研究之后,他们发现这个设计过时了。

[例2] He is *back* to his work.

译文:他回来工作了。

(四) 介词或介词短语转译成动词

[例1] Some people are *for* the operation method, while some are *against* it.

译文:有些人赞成这种操作方法,而有些人反对。

[例2] The worker *in* blue is in charge of loading.

译文:穿蓝色衣服的工人负责装载。

二、转译成名词

(一) 动词转译成名词

英语中很多由名词派生的动词,以及由名词转用的动词,在汉语中往往不易找到相应的动词,这时可将其转译成汉语名词。

[例1] Metals consist of atoms and are *characterized* by metallic bonding.

译文:金属由原子组成,其特征是金属键连接。

[例2] The milling machine is primarily *designed* to produce flat and angular surfaces.

译文:铣床的主要作用是加工平的和倾斜的表面。

(二) 形容词转译成名词

[例 1] Scientists are making an effort in order to put it to *practical* use.

译文:科学家正在努力将其用于实际。

[例 2] Both compounds are acids, the *former* is *strong* and the *latter* is *weak*.

译文:这两种化合物都是酸。前者是强酸,后者是弱酸。

(三) 副词转译成名词

[例] They have not done so well *ideologically*, however, as *organizationally*.

译文:但是,他们的思想工作没有他们的组织工作做得好。

三、转译成形容词

(一) 名词转译成形容词

形容词派生的名词,或者有些名词作表语时,往往可以转译成形容词。

[例 1] The path between points is of little *concern* and is not particularly controlled.

译文:点与点之间的路径并不重要,没有受到特别控制。

[例 2] I realized the great *importance* of solving the problem immediately.

译文:我意识到了立刻解决此问题是非常重要的。

[例 3] The seminar is a great *success*.

译文:这场研讨会非常成功。

(二) 副词转译成形容词

由于英语中的动词在翻译的时候可以转换成汉语名词,所以往往将修饰该动词的副词转译成形容词。

[例 1] The electronic computer is *chiefly* characterized by its accurate and rapid computation.

译文:电子计算机的主要特点是计算迅速、准确。

[例 2] He was *deeply* impressed by what he saw in the factory.

译文:在那个工厂所看到的给他留下了深刻的印象。

四、转译成副词

(一) 名词转译成副词

[例] We found *difficulty* in solving the problem.

译文:我们感到问题难以解决。

(二) 形容词转译成副词

由于英语中的名词在翻译的时候可以转换成汉语动词,所以往往将修饰该名词的形容词转译成汉语副词。

[例 1] A cam movement has been in *practical* use for motion control technology from old times.

译文:很早以来,凸轮的运动实际上已经用于运动控制技术。

[例 2] AC-AC *direct* conversion, not via DC, will be capable of operation with lesser loss under simultaneous current control.

译文:交流电不经过直流电,直接转换成(另一种)交流电,在同步电流控制下,能够使操作过程中的损失减少。

❖ Reading Material

Ceramics, Polymers and Composite Materials

Ceramics are inorganic compounds between metallic and nonmetallic elements; they are most frequently oxides, nitrides, and carbides and silicates. For example, common ceramic materials include aluminum oxide (Al_2O_3, for spark plugs and for substrates for microelectronics), silicon dioxide (SiO_2, for electrical insulation in microelectronics), silicon carbide (SiC, an abrasive), silicon nitride (Si_3N_4, for wear parts), and, in addition, what some refer to as the traditional ceramics—those composed of clay minerals (i. e., porcelain), as well as cement and glass. Ceramics primarily have ionic bonds, but covalent bonding is also present. Usually, they are crystalline in form. Glass is grouped with ceramics because it has similar properties, but most glass is amorphous. With regard to mechanical behavior, ceramic materials are relatively stiff and strong—stiffnesses and strengths are comparable to those of the metals (see Figure 3. 2 and Figure 3. 3). In addition, they are typically very hard. Historically, ceramics have exhibited extreme brittleness (lack of ductility) and are highly susceptible to fracture. However, newer ceramics are being engineered to have improved resistance to fracture; these materials are used for cookware, cutlery, and even automobile engine parts. Furthermore, ceramic materials are typically insulative to the passage of heat and electricity (i. e., have low electrical conductivities), and are more resistant to high temperatures and harsh environments than metals and polymers.

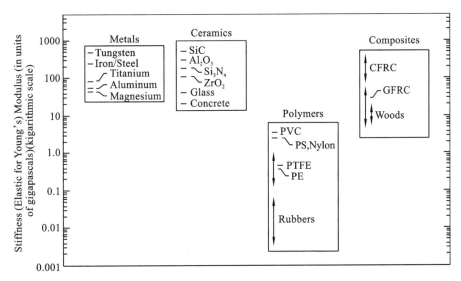

Figure 3. 2 Bar chart of room-temperature stiffness(i. e., elastic modulus) values

Polymers include the familiar plastic and rubber materials. They have very large chainlike molecular structures that often have a backbone of carbon atoms. The polymer chains can be free to slide past one another (thermoplastics) or they can be connected to each other with crosslinks (thermosets or elastomers). Thermoplastics can be reformed and

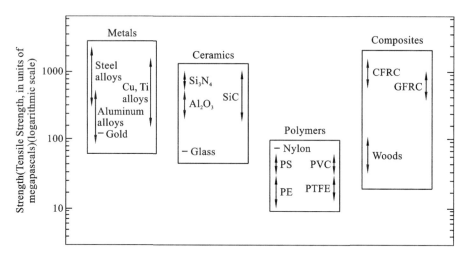

Figure 3.3 Bar chart of room-temperature strength (i. e. , tensile strength) values

recycled, while thermosets and elastomers are not reworkable. Polymer materials typically have low densities, whereas their mechanical characteristics are generally dissimilar to the metallic and ceramic materials—they are not as stiff nor as strong as these other material types. However, on the basis of their low densities, many times their stiffnesses and strengths on a per-mass basis are comparable to the metals and ceramics. In addition, many of the polymers are extremely ductile and pliable (i. e. , plastic), which means they are easily formed into complex shapes. In general, they are relatively inert chemically and unreactive in a large number of environments. One major drawback to the polymers is their tendency to soften and/or decompose at modest temperatures, which, in some instances, limits their use. Furthermore, they have low electrical conductivities and are nonmagnetic. Some of the common and familiar polymers are polyethylene (PE), nylon, polyvinyl chloride (PVC), polycarbonate (PC), polystyrene (PS), and silicone rubber. PVC lends itself to an incredible array of applications, from artificial leather to electrical insulation and cabling, packaging and containers. The versatility of PVC is due to the wide range of plasticisers and other additives that it accepts. The term "additives" in polymer science refers to the chemicals and compounds added to the polymer base to modify its material properties. PE is a cheap, low friction polymer commonly used to make disposable shopping bags and trash bags, and is considered a commodity plastic.

Composite materials are multiphase materials obtained by artificial combination of different materials to attain properties that the individual components cannot attain. An example is a lightweight structural composite, the carbon fiber-reinforced polymer (CFRP) composite, obtained by embedding continuous carbon fibers in one or more orientations in a polymer matrix. The fibers provide the strength and stiffness while the polymer serves as the binder. Thus, CFRP is relatively stiff, strong and flexible. In addition, it has a low density. They are used in some aircraft and aerospace applications, as well as high-tech sporting equipment (e. g. , bicycles, golf clubs, tennis rackets, and skis/snowboards) and recently in automobile bumpers. In general, composites are classified according to their matrix materials.

The main classes of composites are polymer-matrix, cement-matrix, metal-matrix, carbon-matrix, and ceramic-matrix.

Polymer-matrix and cement-matrix composites are the most common due to the low cost of fabrication. In addition to lightweight structures, polymer-matrix composites are used for vibration damping, electronic enclosures, asphalt, and solder replacement. Cement-matrix composites in the form of concrete (with fine and coarse aggregate), or cement paste(without any aggregate) are used for civil structures, prefabricated housing, precast concrete, masonry, landfill cover, thermal insulation, and sound absorption. Carbon-matrix composites are important for lightweight structures(like the space shuttle) and components(such as aircraft brakes) that need to withstand high temperatures, but they are relatively expensive because of the high cost of fabrication. One example is reinforced Carbon-Carbon (RCC), the light gray material which withstands re-entry temperatures up to 1510℃ (2750 ℉) and protects the space shuttle's wing leading edges and nose cap. Ceramic-matrix composites are superior to carbon-matrix composites in oxidation resistance, but they are not as well developed. Metal-matrix composites with aluminum as the matrix are used for lightweight structures and low-thermal-expansion electronic enclosures, but their applications are limited by the high cost of fabrication and by galvanic corrosion.

New Words and Expressions

1. compound   化合物
2. polycarbonate(PC)   聚碳酸酯
3. clay mineral   黏土矿物
4. polystyrene(PS)   聚苯乙烯
5. porcelain   瓷的,瓷料
6. plasticiser   增塑剂
7. cement   水泥
8. additive   添加剂
9. ionic bond   离子键
10. characteristics   特性
11. covalent bonding   共价键结合
12. decompose   分解,(使)腐烂
13. crystalline   结晶质的,晶体
14. multiphase   多相(的)
15. plastic   可塑的,塑料
16. carbon fiber-reinforced polymer   碳纤维增强聚合物
17. rubber   橡胶,橡皮
18. brittleness   脆性
19. orientation   方向,方位,定位
20. cutlery   餐具
21. matrix   基体,矩阵
22. thermoplastics   热塑性塑料
23. vibration damping   减振;振动衰减
24. thermosets   热固性塑料
25. electronic enclosure   电子产品外壳
26. elastomer   弹性体,合成橡胶
27. asphalt   沥青,柏油
28. nylon   尼龙
29. solder   焊料,(低温)焊接
30. polyethylene (PE)   聚乙烯
31. galvanic corrosion   电(化腐)蚀,电偶腐蚀
32. polyvinyl chloride(PVC)   聚氯乙烯
33. GFRP   玻璃纤维增强塑料

Lesson 4 Mechanical Design

Text

Mechanical design is the application of science and technology to devise new or important products for the purpose of satisfying human needs. It is a vast field of engineering technology which not only concerns itself with the original conception of the product in terms of its size, shape and construction details, but also considers the various factors involved in the manufacture, marketing and use of the product.

A machine is a combination of mechanisms and other components which transforms, transmits, or utilizes energy, force, or motion for a useful purpose. The term "mechanical design" is used in a broader sense than "machine design" to include their design. For some apparatus, the thermal and fluid aspects that determine the requirements of heat, flow path, and volume are separately considered. However, the motion and structural aspects and the provisions for retention and enclosure are considerations in mechanical design. Applications occur in the field of mechanical engineering, and in other engineering fields as well, all of which require mechanical devices, such as switches, cams, valves, vessels, and mixers.

Mechanical design covers the following contents:

(1) Providing an introduction to the design process, problem formulation, safety factors.

(2) Reviewing the material properties and static and dynamic loading analysis, including beam, vibration and impact loading.

(3) Reviewing the fundamentals of stress and defection analysis.

(4) Introducing static failure theories and fracture-mechanics analysis for static loads.

(5) Introducing fatigue-failure theory with the emphasis on stress-life approaches to high-cycle fatigue design, which is commonly used in the design of rotation machinery.

(6) Discussing thoroughly the phenomena of wear mechanisms, surface contact stresses, and surface fatigue.

(7) Investigating shaft design using the fatigue-analysis techniques.

(8) Discussing fluid-film and rolling-element bearing theory and application.

(9) Giving a thorough introduction to the kinematics, design and stress analysis of spur gears, and a simple introduction to helical, bevel, and worm gearing.

(10) Discussing spring design including helical compression, extension and torsion springs.

(11) Dealing with screws and fasteners including power screw and preload fasteners.

(12) Introducing the design and specification of disk and drum clutches and brakes.

1. The Design Process

Designing starts with a need, real or imagined. Existing apparatus may need improvements in durability, efficiency, weight, speed, or cost. New apparatus may be needed to perform a function previously done by men, such as computation, assembly, or servicing.

Design is an iterative process. The starting point is a market need or a new idea; the end point is the full specifications of a product that fills the need or embodies the idea. It is essential to define the need precisely, that is, to formulate a need statement, often in the form: a device is required to perform task "X". Writers on design emphasize that the statement should be solution-neutral (that is, it should not imply how the task will be done), to avoid narrow thinking limited by pre-conceptions. Between the need statement and the product specification lie the set of stages shown in Figure 4.1: The stages of conceptual design, embodiment design and detailed design.

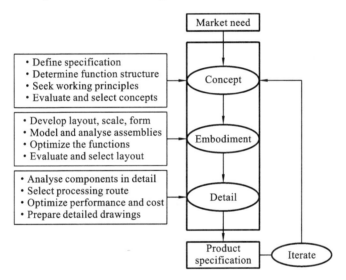

Figure 4.1 The design flow chart

The product itself is called a technical system. A technical system consists of assemblies or subassemblies and components, put together in a way that performs the required task, as in the breakdown of Figure 4.2. It is like describing a cat (the system) as made up of one head, one body, one tail, four legs, etc. (the assemblies), each composed of components—femurs, quadriceps, claws, fur. This decomposition is a useful way to analyze an existing design, but it is not of much help in the design process itself, that is, in the synthesis of new designs. Better, for this purpose, is one based on the ideas of systems analysis; it thinks of the inputs, flows and outputs of information, energy and materials, as in Figure 4.3. The design converts the inputs into the outputs. An electric motor converts electrical into mechanical energy; a forging press takes and reshapes material; a burglar alarm collects information and converts it to noise. In this approach, the system is broken down into connected subsystems which perform specific sub-functions, as in Figure 4.3; The resulting arrangement is called the function structure or function decomposition of the system. It is like describing a cat as an appropriate linkage of a respiratory system, a cardio-vascular system, a nervous system, a digestive system and so on. Alternative designs link the unit functions in alternative ways, combine functions, or split them. The function-structure gives a systematic way of assessing design options.

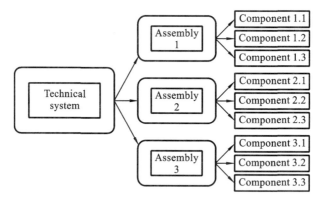

Figure 4.2　The analysis of a technical system as a breakdown into assemblies and components

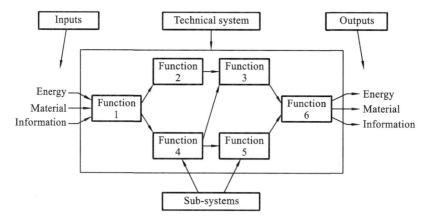

Figure 4.3　The systems approach to the analysis of a technical system

The design proceeds by developing concepts to fill each of the sub-functions in the function structure, each based on a working principle. At this, the conceptual design stage (see Figure 4.1 again), all options are open: the designer considers alternative concepts for the sub-functions and the ways in which these might be separated or combined. The next stage, embodiment, takes each promising concept and seeks to analyse its operation at an approximate level, sizing the components, and selecting materials which will perform properly in the ranges of stress, temperature and environment suggested by the analysis or required by the specification, examining the implications for performance and cost. The embodiment stage ends with a feasible layout which is passed to the detailed design stage. Here specifications for each component are drawn up; critical components may be subjected to precise mechanical or thermal analysis; optimization methods are applied to components and groups of components to maximize performance; a final choice of geometry and material is made, the production is analysed and the design is costed. The stage ends with detailed production specifications.

2. Some rules for design

To stimulate creative thought, the following rules are suggested for the designer and analyst. The first six rules are particularly applicable for the analyst, although he may become involved with all the rules.

(1) Apply ingenuity to utilize desired physical properties and to control undesired ones.

(2) Recognize functional loads and their significance.

(3) Anticipate unintentional loads.

(4) Devise more favorable loading conditions.

(5) Provide for favorable stress distribution and stiffness with minimum weight.

(6) Use basic equations to proportion and optimize dimensions.

(7) Choose materials for a combination of properties.

(8) Select carefully between stock and integral components.

(9) Modify a functional design to fit the manufacturing process and reduce cost.

(10) Provide for accurate location and noninterference of parts in assembly.

New Words and Expressions

1. mechanism 机构,机械
2. thermal 热量;热的
3. switch 开关
4. cam 凸轮
5. valve 阀门
6. beam 梁
7. phenomena 现象
8. screw 螺钉,螺杆
9. fasteners 紧固件
10. spring 弹簧
11. gear 齿轮
12. durability 耐用性
13. femur 股骨
14. quadriceps 四头肌
15. optimization method 最优化方法
16. stiffness 硬度
17. stock 原料,备品
18. noninterference 互不干扰

Notes

(1) However, the motion and structural aspects and the provisions for retention and enclosure are considerations in mechanical design. Applications occur in the field of mechanical engineering, and in other engineering fields as well, all of which require mechanical devices, such as switches, cams, valves, vessels, and mixers.

但是,在机械设计时要考虑运动和结构方面的问题及保存和封装的规定。机械设计机械工程领域以及其他工程领域都有应用,这些领域都需要诸如开关、凸轮、阀门、容器和搅拌器之类的机械装置。

(2) This decomposition is a useful way to analyze an existing design, but it is not of much help in the design process itself, that is, in the synthesis of new designs. Better, for this

purpose, is one based on the ideas of systems analysis; it thinks of the inputs, flows and outputs of information, energy and materials.

这种分解是分析现有设计的有用方式,但对设计过程本身即新设计的综合分析没有太大的帮助。因此,最好是采用基于系统分析思想的方法,这种方法涉及信息、能量、材料的输入、流动和输出。

❖ Translation Skills

词量的增减

词汇的准确翻译绝不意味着形式上保持词量的相等,不允许增减一些词。相反,翻译时允许改变词量,而且常常是必须改变词量。这种增减一些原文中没有的词或是减去原文中某些词的译法,叫词量增减。

一、词量增加

(一) 重复英语省略的某些词

[例] Some of the gases in the air are constant in amount, while others are not.

译文:空气中有些气体的含量相当稳定,但有些不稳定。

(二) 增加关联词语

[例] White or shining surfaces reflect heat; dark surfaces absorb it.

译文:白色的或明亮的表面反射热,而黑暗的表面吸收热。

(三) 增加英语复数表达的概念

[例] However, in spite of all this similarity between a voltmeter and an ammeter there are also important differences.

译文:尽管伏特表和安培表之间有这些类似之处,然而它们之间还有若干重要差别。

(四) 英语抽象名词的增译

[例] These principles will be illustrated by the following transition.

译文:这些原理将由下列演变过程来说明。

(五) 逻辑加词

[例1] Air pressure decreased with altitude.

译文:气压随海拔高度的增加而下降。

[例2] It may well improve long term weather forecasting.

译文:这会提高长期天气预报的准确率。

(六) 修辞加词

[例] This question is really a circuit design rather than a layout problem.

译文:这个问题实际上是属于线路设计方面的而不是线路布局方面的问题。

(七) 汉语中补充概括总结性词语

[例] The resistance of the pipe to the flow of water through it depends upon the length of the pipe, the diameter of the pipe, and the mature of the inside walls (rough or smooth).

译文:水管对流过的水的阻力取决于下列三个因素:管长、管道直径、管道内壁的性状(粗糙或光滑)。

（八）时态加词

[例] Though some of the most brilliant minds of history attacked the problem, the dream was not realized until one cold December morning in 1903 when two American brothers rose from the sands of North Carolina to bring wings to man.

译文：虽然历史上有过一些最有才华的人曾经致力于解决这个难题，但这个梦想直到 1903 年 12 月一个寒冷的早晨才得以实现。当时，美国有对兄弟从北卡莱罗纳州的沙滩上腾空而起，给人类添上了双翼。

二、词量减少

（一）省略冠词

[例 1] Any substance is made up of atoms whether it is *a* solid, *a* liquid, or *a* gas.

译文：任何物质，不管它是固体、液体或气体，都是由原子组成的。（省略三个不定冠词 a）

[例 2] As *the* crankshaft turns to *the* left and upward, *the* connecting rod pushes *the* piston upward.

译文：当曲轴向左上方转动时，连杆把活塞推向上方。（省略四个定冠词 the）

（二）省略代词

[例 1] Man has always been interested in extending the range of *his* senses and the power of his mind.

译文：人类一向对扩大感觉范围和提高智力有兴趣。（省略物主代词 his）

[例 2] A gas distributed *itself* uniformly through a container.

译文：气体均匀地分布在整个容器中。（省略反身代词 itself）

（三）省略介词

[例 1] *In* winter, it is much colder *in* the North than it is *in* the South.

译文：冬天，北方比南方冷得多。（省略三个介词 in）

[例 2] The light *during* the day is given by the sun.

译文：白天的亮光太阳发出的。（省略介词 during）

[例 3] The air was removed *from* between the two pipes.

译文：两根管子之间的空气已抽出。（省略介词 from）

（四）省略连接词

[例 1] Like charges repel each other *while* opposite charges attract.

译文：同性电荷相斥，异性电荷相吸。（省略连词 while）

[例 2] The density of a body can be found *providing* its mass and volume are known.

译文：已知物体的质量和体积就可以求出其密度。（省略连词 providing）

（五）省略动词

1. 省略谓语动词

英语的谓语必须有动词，而谓语又是句中不可缺少的成分，所以英语的句子离不开动词。但汉语则不是，汉语可以直接用形容词、名词、介词结构、主谓结构等作谓语。翻译时，可以省略谓语动词，尤其是系动词。

[例 1] A square *has* four equal sides.

译文：正方形四边相等。（省略动词 has）

[例 2] The nucleus of the U-235 *is* easy to break.

译文：铀235原子核容易破裂。（省略动词is）

［例3］ This laser beam *covers* a very narrow range of frequencies.

译文：该激光束的频率范围很窄。（省略动词cover）

2. 省略与具有动作意义的名词连用的动词

科技英语中往往习惯使用具有动作意义的名词来表示句中的主要动作意义，而用不表示实际动作意义的动词作谓语。翻译时，可以把具有主要动作意义的名词译为谓语，而省略与其连用的动词。

［例1］ In conduction and convection energy *transfer* through a material medium *is involved*.

译文：在传导和对流时，能量通过某种材料介质传递。（省略动词involve）

［例2］ Energy *losses* due to friction *occur* in every machine.

译文：每台机器都由于摩擦而损耗能量。（省略动词occur）

（六）省略名词

1. 省略 of 前的名词

［例1］ *The number* of known hydrocarbons runs into tens of thousands.

译文：已知的碳氢化合物有几万种。

［例2］ Going through *the process* of heat treatment, metals become much stronger and more durable.

译文：金属经过热处理后，强度更大，更加耐用。

2. 省略同义的名词

［例］ The essentials of an electric *generator or dynamo* are powerful magnets and a rapidly moving coil.

译文：发电机的主要部件是强力的磁铁和高速运动的线圈。

❖ Reading Material

Machine Design

The complete design of a machine is a complex process. The designer must have a good background in such fields as statics, kinematics, dynamics, and strength of materials, and in addition, be familiar with fabricating materials and processes. The designer must be able to assemble all the relevant facts, and make calculations, sketches, and drawings to convey manufacturing information to the shop.

One of the first steps in the design of any product is to select the material from which each part is to be made. Numerous materials are available to today's designers. The function of the product, its appearance, the cost of the material, and the cost of fabrication are important in making a selection. A careful evaluation of the properties of a material must be made prior to any calculations.

Careful calculations are necessary to ensure the validity of a design. Calculations never appear on drawings, but are filed away for several reasons. In case of any part failures, it is desirable to know what was done in originally designing the defective components. Also, an experience file can result from having calculations from past projects. When a similar design is needed, past records are of great help.

The checking of calculations (and drawing dimensions) is of utmost importance. The misplacement of one decimal point can ruin an otherwise acceptable project. For example, if one were to design a bracket to support 100 lb when it should have been figured for 1000 lb, failure would surely be forthcoming. All aspects of design work should be checked and rechecked.

The computer is a tool helpful to mechanical designers to lighten tedious calculations, and provide extended analysis of available data. Interactive systems, based on computer capabilities, have made possible the concepts of computer-aided design (CAD) and computer-aided manufacturing (CAM). Through such systems, it is possible for one to transmit conceptual ideas to punched tapes for numerical machine control without having formal working drawings.

Laboratory tests, models, and prototypes help considerably in machine design. Laboratories furnish much of the information needed to establish basic concepts; however, they can also be used to gain some idea of how a product will perform in the field.

Finally, a successful designer does all he can to keep up to date. New materials and production methods appear daily. Drafting and design personnel may lose their usefulness by not being versed in modern methods and materials. A good designer reads technical periodicals constantly to keep abreast of new developments.

New Words and Expressions

1. statics 静力学
2. kinematics 运动学
3. dynamics 动力学
4. strength of materials 材料强度
5. fabricate 装配
6. relevant 有关的
7. calculation 计算
8. sketch 草图
9. numerous 很多的
10. evaluation 估价
11. property 特性
12. defective 有缺陷的
13. component 组成部分
14. result from 产生于……
15. utmost 极度的
16. misplacement 误放
17. decimal point 小数点
18. bracket 托架
19. forthcome 在前面出现
20. prototype 原型
21. keep up to date 跟上时代
22. personnel 员工(总称)
23. periodical 期刊
24. lb 磅,1磅＝0.45359237千克

Lesson 5 Machinery Components

Text

1. Shafts

Shafts are indispensable mechanical power transmission elements that are mainly used for transmitting rotary motion and torque from one point to another point (see Figure 5.1). Elements such as gears, pulleys (sheaves), flywheels, clutches, and sprockets are mounted on the shaft and are used to transmit power from the motor or engine to a machine. The rotational force (torque) is transmitted to these elements on the shaft by press fit, keys, splines and pins. The shaft rotates on rolling contact or bush bearings. Various types of retaining rings, thrust bearings, grooves and steps in the shaft are used to take up axial loads and locate the rotating elements. Since shafts take on several different configurations and are used for many different purposes, several different definitions are in common use, as listed below.

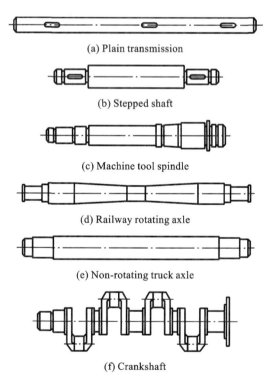

(a) Plain transmission

(b) Stepped shaft

(c) Machine tool spindle

(d) Railway rotating axle

(e) Non-rotating truck axle

(f) Crankshaft

Figure 5.1 Various shaft types

Solid shafts   A solid shaft is obtainable commercially in round bar stock up to 15 cm in diameter. It is produced by hot-drawing and cold-drawing or by machining-finishing with diameters in increments of 6 mm or less. For large sizes, special rolling procedures are required, and for extremely large shafts, billets are forged to the proper shape. Particularly in a solid shafting, the shaft is stepped to allow greater strength in the middle portion with minimum diameter on the ends at the bearings. The steps allow shoulders for positioning the various parts pressed onto the shaft during the rotor assembly.

Hollow shafts To minimize weights, solid shafts are bored out or drilled, or hollow pipes and tubing are used. Hollow shafts also allow internal support or permit other shafts to operate through the interior. A hollow shaft, to have the same strength in bending and torsion, has a larger diameter than a solid shaft, but its weight is less. The centers of larger shafts made from ingots are often bored out to remove imperfections and also to allow visual inspection for forging cracks.

Axles An axle is a supporting shaft on or with which a wheel or a set of wheels revolve. Axles normally carry only transverse loads, but occasionally, they also transmit torsion loads. The wheel or gear can be attached to the axle with a built in bearing or bushing. The purpose of an axle is to secure the wheels to specific locations relative to other wheels.

Spindles A spindle is a short, rotating shaft that serves as axes for larger rotating parts. Spindles are essential in all kinds of manufacturing. They are used to perform various tasks in a machine tool like grinding, milling, drilling, turning, engraving and routing. High speed spindles are designed with the spindle motor flanged to the spindle shaft, while conventional or normal spindles are mostly gear or belt driven.

Shafts design and manufacturing Axial dimensions are often fixed from the layout of the mechanism. Shaft design is to determine the diameter of the shaft such that it withstands the applied loads, after stress concentrations, with a known factor of safety. The material used for ordinary commercial shafting is mid steel. For high strength alloy steel such as, nickel, nickel chromium or chrome-vanadium steel is used. Shafts are generally formed by hot rolling and finished to the required size by cold drawing or turning and grinding. The cold rolled shafts are stronger than hot rolled shafts but with higher residual stresses. The residual stresses may cause distortion of the shaft when it is machined, especially when slots or key ways are cut. Shafts of larger diameter are usually forged and turned to size in a lathe.

2. Shaft accessories

1) Keys

A key is a machinery component inserted into the keyway of two mating parts to prevent relative angular or sliding motion between the parts (see Figure 5.2). It is mainly employed to transmit torque from a shaft to a hub or vice versa.

Parallel keys Both the square and the rectangular keys are referred to as parallel keys because the top, bottom and the sides of the keys are parallel. This is the most standardized version of keys currently being used in all applications. The parallel key is installed on the shaft, before mating the hub. The hub is slide over the key to provide the interface.

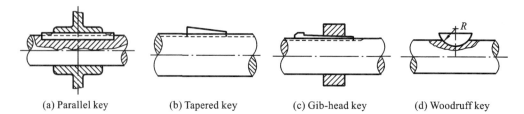

(a) Parallel key   (b) Tapered key   (c) Gib-head key   (d) Woodruff key

Figure 5.2 Various key types

Tapered keys The tapered keys have a rectangular varying cross section. The basic idea of using tapered keys is mainly for the ease of assembly and removal of the key. Further due to the tapered section of the key along its length, it acts as an axial retainer for the mechanical power transmitting element. A tapered key is installed after mating the hub and shaft. The taper extends over the length of the hub.

Gib-Head Keys This type of key is quite similar to the tapered key except for a head which is provided at the bigger end of the key in order to facilitate easy removal of the key. This type of key is used in areas where frequent removal of the key is necessary. The gib-head key acts to prevent relative axial motion. This also gives the advantage that the hub position can be adjusted for the best axial location.

Woodruff Keys The woodruff key is so called because of its manufacturing method. The key is in the shape of a section of a circle. This type of key tends to have a lesser shear strength compared to that of a rectangular or square key. Woodruff keys are used in light loading applications, where it is desirable to easy assembly and disassembly. A Woodruff key also yields better concentricity after assembly of the wheel and shaft. This is especially important at high speeds, as, for example, with a turbine wheel and shaft.

The type and size of the key is usually selected after the shaft and hub have been designed. The length and material specification is determined through design analysis, while two factors limit their selection: width of the hub and distance to adjacent stress concentrations. Keys are commonly made of low-carbon, cold-drawn steel; such as AISI 1020CD with 61 ksi ultimate strength. A key is normally held in place by: set screws, shoulders, retaining rings, or spacers.

2) Splines

A spline can be described as a series of axial keys machined into a shaft(see Figure 5.3), with corresponding grooves machined into the bore of the mating part (gear, sheave, sprocket, etc). The splined shafts are used when the force to be transmitted is large in proportion to the size of the shaft as in automobile transmission and sliding gear transmissions. There are two major types of splines used in industry: (1) straight-sided splines, and (2) involute splines. The

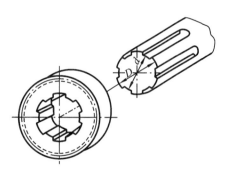

Figure 5.3 Typical spline

latter are stronger, more easily measured, and have a self-centering action when twisted. Splines are either "cut" (machined) or rolled. Rolled splines are stronger than cut splines due to the cold working of the metal. Nitriding is common to achieve very hard surfaces which reduce wear.

3) Pins

With the element in position on the shaft, a hole can be drilled through both the hub and the shaft, and a pin can be inserted in the hole. Tapered circular pins can be used to retain shaft-mounted members from both axial and rotational movement (see Figure 5.4), while the straight, solid, cylindrical pin is subjected to shear over two cross sections. One problem with cylindrical pins is that fitting them

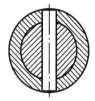

Figure 5.4 Taper pin

adequately to provide precise location of the hub and to prevent the pin from falling out is difficult. The taper pin overcomes some of these problems, as does the split spring pin. Sometimes the diameter of the pin is purposely made small to ensure that the pin will break if a moderate overload is encountered, in order to protect critical parts of a mechanism. Such a pin is called a shear pin.

4) Couplings

The term coupling refers to a device used to connect two shafts together at their ends for the purpose of transmitting power. There two general types of couplings: rigid and flexible.

Rigid Couplings As shown in Figure 5.5 (a), rigid couplings are designed to draw two shafts together tightly so that no relative motion can occur between them. Rigid couplings can be classified as: sleeve couplings, split couplings and flange couplings. Rigid couplings should be used only when the alignment of the two shafts can be maintained very accurately, not only at the time of installation but also during operation of the machines. If significant angular, radial, or axial misalignment occurs, stresses that are difficult to predict and that may lead to early failure due to fatigue will be induced in the shafts. These difficulties can be overcome by the use of flexible couplings.

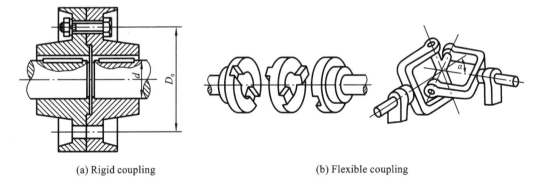

(a) Rigid coupling   (b) Flexible coupling

Figure 5.5 Typical couplings

Flexible Couplings Flexible couplings are designed to transmit torque smoothly while permitting some axial, radial, and angular misalignment. The flexibility is such that when misalignment does occur, parts of the coupling move with little or no resistance. Thus, no significant axial or bending stresses are developed in the shaft. According to the elastic elements, flexible couplings can be classified as flexible couplings without elastic elements and with elastic elements. The former include: cross-sliding couplings, slider couplings, universal joints coupling, gear couplings and double roller chain couplings. Flexible couplings with elastic elements can be classified as pin couplings with elastic sleeves, elastic pin couplings, couplings with elastic spider, couplings with rubber type element and diaphragm couplings.

When selecting a specific type of coupling, several factors should be considered: (1) Torque-carrying capacity; (2) Axial movement of one or both shafts during operation; (3) Shaft alignment; (4) Vibration control; (5) Torsion deflection; and (6) Operating life and price.

New Words and Expressions

1. pulley　滑轮，带轮
2. torque　扭矩，转（力）矩
3. sheave　滑轮车，槽轮
4. disassembly　拆卸，分解
5. stock　棒料，库存
6. woodruff key　半圆键，月牙键
7. axle　轮轴，车轴
8. spline　花键；用花键连接
9. bushing　轴瓦，轴衬
10. involute spline　渐开线花键
11. spindle　主轴，轴
12. groove　沟，槽；刻沟，刻槽
13. residual stress　残余应力
14. coupling　联轴器
15. distortion　变形，挠曲
16. misalignment　未对准（线），非对中
17. referred to as　把……称作，把……当做

Notes

（1）The shaft rotates on rolling contact or bush bearings. Various types of retaining rings, thrust bearings, grooves and steps in the shaft are used to take up axial loads and locate the rotating elements.

传动轴一般用滚动或滑动轴承支承，用推力轴承、键槽、轴肩和定位环等为旋转零部件定位并且承受轴向载荷。

（2）A solid shaft is obtainable commercially in round bar stock up to 15 cm in diameter. It is produced by hot-drawing and cold-drawing or by machining-finishing with diameters in increments of 6 mm or less.

实心轴可利用直接购买的直径最大可达到 15 cm 的棒料通过热拔、冷拔或机加工制成，轴的直径可由 6 mm 或更小开始递增。

（3）High speed spindles are designed with the spindle motor flanged to the spindle shaft, while conventional or normal spindles are mostly gear or belt driven.

高速主轴一般通过法兰与主轴电动机连接，而普通主轴常常通过齿轮或传动带驱动。

(4) The cold rolled shafts are stronger than hot rolled shafts but with higher residual stresses. The residual stresses may cause distortion of the shaft when it is machined, especially when slots or key ways are cut.

冷轧加工的传动轴强度比热轧加工的高,但会存在较大的残余应力。在对传动轴进行机加工特别是车槽或加工键槽时,残余应力的存在会导致轴发生挠曲变形。

(5) Both the square and the rectangular keys are referred to as parallel keys because the top, bottom and the sides of the keys are parallel.

方键和矩形键都是平键,其上下、左右侧面互相平行。

(6) Rigid couplings should be used only when the alignment of the two shafts can be maintained very accurately, not only at the time of installation but also during operation of the machines.

刚性联轴器只能用于连接在安装和使用过程中对中性都很好的两根转动轴。

❖ Translation Skills

词义的选择和引申

一、词义的选择

科技英语中一词多义的现象非常普遍,例如:case 一词在不同的学科专业领域中有不同的词义,如病例、案例、机器壳体等。因此,把握词汇的本质词义是选择词义的基础。译者应避免脱离本质词义的"臆造",但正确选择词义并不意味着照搬字典,而是根据上下文内容反复推敲,把握原文的精神实质,选择最确切的词义。词义的选择一般从以下几方面考虑。

(一) 根据学科专业领域选择词义

在选择专业词汇的词义时,首先要了解所译文献学科专业领域,尽量使用权威性专业词典,尽量采用国家技术标准中的术语,避免采用非标准专业术语或俗语。

[例 1] *Milling* is *a machining process for* removing material by relative motion between *a workpiece* and *a rotating cutter* having multiple *cutting edges*.

译文:铣削是一种通过工件与多刃旋转铣刀之间的相对运动去除多余材料的加工工艺。

[例 2] The thickness of a *tooth* measured along the *pitch circle* is one half the *circular pitch*.

译文:沿(齿轮的)节圆所测得的齿厚是周节的二分之一。

(二) 根据汉语习惯选择词义

普通词汇和专业词汇一样,也存在着词义选择问题,如果词义选择不当,会使译文难以理解。因此,对于所译文献中的普通词汇,在选择词义时,应使用规范的和符合汉语语言习惯的词义。

[例 1] Noises may *develop* in a worn engine.

译文:磨损了的发动机可能会产生噪声。

[例 2] Other isolation methods are being *developed*.

译文:目前正在研究其他的隔离方法。

二、词义引申

在科技英语翻译过程中,当英语句子中的某个词或词组按词典中的释义直接译出,不符合汉语的修辞习惯或语言规范时,可以结合上下文,根据语气、逻辑关系、搭配习惯及全句的技术含义等方面的情况,对词义进行引申。引申翻译应遵循以下原则:(1)基于词义引申;(2)根据上下文的逻辑关系进行引申;(3)符合汉语的搭配习惯;(4)符合专业技术规范。

(一) 引申分类

科技英语翻译时,从语法的角度,要注意动词、名词、形容词和词组之间的搭配关系。遇到不符合汉语搭配习惯的情况时,可采用引申翻译。

1. 动词意义的引申

[例1] When sunlight *falls* on the leaves of plants, it is transformed into chemical energy.

译文:当阳光照射在植物的叶片上时,光能即转化为化学能。(直译:落到)

[例2] The oil also *provides* some cooling effect.

译文:润滑油也起到一定的冷却作用。(直译:提供)

2. 名词意义的引申

[例1] Other *things* being equal iron heats up fast than aluminum.

译文:其他条件相同时,金属铁比铝热得快。(直译:情况)

[例2] Laser is one of the most sensational *developments* in recent years.

译文:激光是近年来轰动一时的科学成就之一。(直译:发展)

3. 形容词意义的引申

[例1] Such particles are far too tiny to be seen with the *strongest* microscope.

译文:这些粒子太小,即使用最高倍的显微镜也分辨不出来。(直译:最强大的)

[例2] The sun's heat offers an almost *limitless* source of power.

译文:太阳能提供了一个几乎取之不尽的动力源泉。(直译:无限的)

4. 词组意义的引申

[例1] Rubber is not hard, it *gives way to pressure*.

译文:橡胶质软,受压就会变形。(直译:为压力让路)

[例2] Solar energy seems to *offer more hope* than any other source of energy.

译文:太阳能似乎比其他能源更有前途。(直译:提供更多希望)

(二) 引申方法

1. 技术引申

技术性引申的目的是使译文中涉及的科学技术概念的词义更符合汉语语言规范。

[例1] After the spring has been closed to its *solid height*, the compressive force is removed.

译文:弹簧被压缩到接近并紧高度之后,就没有了压力。(直译:固体高度)

[例2] That engine sounds *good*.

译文:那台发动机运转正常。(直译:良好的)

2. 修辞性引申

为了使译文的语言更加流畅、文句更加通顺,有时需要进行修辞性引申。

[例] This machine is *simple in design*, yet it is efficient in operation.

译文:这台机器结构简单,而且运转效率高。(直译:设计简单)

3. 具体化引申

当英语句子中具有含义比较笼统、抽象的词或词组时，如果直接译出会造成译文概念不清，或用语不符合汉语习惯时，可进行具体化引申。

［例1］ When this assumption is *made*, the vessel is called a thin-walled pressure vessel.

译文：当这一假设成立时，该容器称为薄壁压力容器。（直译：做出）

［例2］ There is a wide area of *performance duplication* between numerical control and automatics.

译文：数控和自动化机床有很多相同的性能。（直译：性能复制）

4. 抽象化引申

依据汉语的表达习惯，将原文中不需强调的具体词义引申为较抽象的词或词组，或把词义较形象的词或词组引申为词义较一般的词或词组，能使译文的结构更紧凑，语言的连贯性更强，且通俗易懂。

［例1］ Steel and cast iron also differ in *carbon*.

译文：钢和铸铁的含碳量不相同。（直译：碳）

［例2］ The facts have been set down *in black and white*.

译文：这些事实已经被清楚地记录下来了。（直译：用黑白色）

❖ Reading Material

Clutches, Set Screws and Springs

1. Clutches

Clutches are employed to connect and disconnect shafts during their relative motion (under load) or at standstill. Shafts and other parts linked by clutches should be strictly collinear. Even slight misalignment deteriorates the performance of clutch and leads to their rapid failure. Clutches are classified into following ways.

Friction clutches The principle of a friction clutch is based on the developing of friction forces between the members of the clutch. Friction force can be regulated by varying the force with which the rubbing surfaces (see Figure 5.6 (a)) are pressed together. Smooth

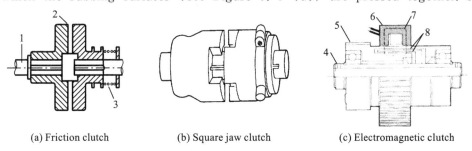

(a) Friction clutch (b) Square jaw clutch (c) Electromagnetic clutch

Figure 5.6　Typical clutches

1—dring shaft; 2—friction plate; 3—spring; 4—output shaft;
5—input rotor; 6—stator; 7—flux path; 8—powder cavity

engagement of clutches is import to avoid high dynamic loads and noise in starting a machine. During the period of engagement (period of acceleration) and disengagement of the clutch, slipping may occur in the clutch. At steady motion, there is no slipping. Incidental slipping may occur at peak loads. With respect to the shape of their friction surfaces, friction clutches can be classified as: disk clutches, cone clutches, and radial clutches.

Positive contact clutches Jaw or toothed clutches (see Figure 5.6 (b)) are employed to transmit considerable torque under conditions of small overall size, in frequent engagements and when smooth engagement is not compulsory. These clutches provide positive engagement and that is why they are called positive clutches.

Electromagnetic clutches The principle of this clutch is based on the fact that the iron powder in the clutch, when magnetized by the magnetic flux, resists shear, and its shear strength increases with the intensity of magnetization. The greatest relative displacement of the particles of powder occurs at the middle of the layer. The layers of powder adjacent to the magnetized surfaces do not move with respect to these surfaces and are not subject to wear. Advantages are: (1) exceptionally rapid action; (2) possibility of fine control of the torque; (3) high wear resistance. These clutches operate with their volume rather than their surfaces and can transmit large torque.

2. Set screws

A set screw is a threaded fastener driven radically through a hub to bear on the outer surface of a shaft. With the exception of the antiquated square head, set screws are headless fasteners and therefore threaded for their entire length. In addition, set screws have a specified point type. The point is used to provide various amounts of holding power when used. The available point types for set screws are the cone, cup, flat, oval, and dog (full or half) points (see Figure 5.7). The point bears on the shaft or digs slightly into its surface. The set screw transmits torque by the friction between the point and the shaft or by the resistance of the material in shear. The capacity for torque transmission is somewhat variable, depending on the hardness of the shaft material and the clamping force created when the screw is installed.

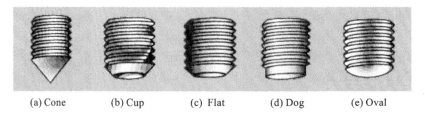

(a) Cone   (b) Cup   (c) Flat   (d) Dog   (e) Oval

Figure 5.7 Various set screw point types

A cup point screw is the most commonly used set screw featuring a cup-shaped indentation on one end. The cup point of this screw is mainly used for a quick, semi-permanent or permanent part assembly where the cutting in the cup point edge of the screw is acceptable. A flat point screw has a flat surface on one end and is indeed the simplest and least expensive type of set screw available. This type of screw would be most appropriate when you need the set screw to press completely flat against a surface in order to create compression. A flat point set screw is also excellent for projects that require frequent relocation, rather than permanent positioning. A pointed tip on one end gives a cone point set screw its name. It is used for the permanent setting of parts. These set screws offer the highest holding capability, due to the deep penetration of the cone point. An oval point set screw features an oval shaped point on one end. This set screw is used in an environment where frequent adjustments are needed and minimal deforming of the part is required. The half dog point screw has a flat tip protruding from one end which is normally located into a groove in a shaft and allows the shaft to rotate whilst retaining the part in place. This type of screw is typically used for a permanent setting of parts and often is used in place of a dowel pin.

3. Springs

A spring is a load-sensitive, energy-storing device, the chief characteristics of which are an ability to tolerate large deflections without failure and to recover its initial size and shape when loads are removed. Although most springs are mechanical and derive their effectiveness from the flexibility inherent in metallic elements, hydraulic springs and air springs are also obtainable. Springs are used for a variety of purposes, such as supplying the motive power in clocks and watches, cushioning transport vehicles, measuring weights, restraining machine elements, making resilient connections, launching and retarding missiles and vehicles, mitigating the transmission of periodic disturbing forces from unbalanced rotating machine to the supporting structure, and providing shock protection for delicate instruments during shipment.

Mechanical springs Mechanical springs may be classified as either wire springs, flat springs, or special-shaped springs, and there are variations within these divisions. Wire springs include helical springs of round or square wire that are cylindrical or conical in shape and are made to resist tensile, compressive, or torsion loads. Under flat springs are included the cantilever and elliptical type (leaf) springs (see Figure 5.8), the wound motor or clock type power springs and the flat spring washers, usually called Belleville springs.

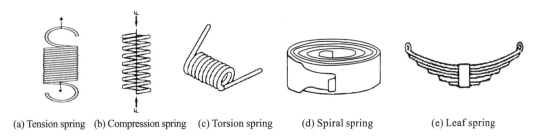

(a) Tension spring (b) Compression spring (c) Torsion spring (d) Spiral spring (e) Leaf spring

Figure 5.8 Various spring types

Air and hydraulic springs An air spring is basically a column of air confined with a rubber-and-fabric container that may take the form of a bellows. The spring action results from the compression and expression and expansion of the air. Air springs provide means for control of load capacity, height, and spring rate; they are used in aircraft landing gear, shock and vibration isolator, seat suspensions, and for suspension systems for machines and vehicles. Hydraulic or liquid springs are comparatively small, thick-walled, hydraulic cylinders in which the spring effect is produced by applying a load to the fluid in the cylinder through a comparably small piton entering at the center of one end of the cylinder. The piston movement, which represents the spring deflection, results from the compression of the fluid and the deformation (bulging) of the cylinder. Some special oils, under a given pressure, have twice the percentage of volume reduction of water.

A helical spring and a hydraulic spring of the same energy capacity are approximately equal in weight; the hydraulic spring, however, is much smaller. Hydraulic springs are particularly suited for applications in which high load capacity and stiffness are required. They are widely used in punch and die assemblies in the machine tools and as recoil springs and buffers in gun systems.

New Words and Expressions

1. clutch 离合器,把握,抓紧
2. indentation 凹口,缺口
3. standstill 静止状态,停顿
4. protrude 伸出,突出
5. engagement 啮合
6. resilient 能复原的,有弹性的
7. cantilever 悬臂,悬臂梁
8. kinematic viscosity 动力黏度
9. bulging 膨胀,褶皱
10. set screw 定位螺钉
11. recoil 反冲;返回,退缩
12. dowel pin 定位销,开槽销
13. buffer 缓冲器;缓冲,减轻
14. Belleville spring 盘形(贝氏)弹簧
15. positive contact clutch 刚性离合器

Lesson 6　The Basic Components of an Automobile

Text

The automobile industry is a fast developing industry. Now there are thousands of factories all over the world manufacturing numerous types of automobile. There is a great demand for varied types of automotive products, vehicles and engines. Today's automobile contains more than 15,000 separate, individual parts that must work together. Some of them make it more comfortable or better looking, but most of them are to make it run. These parts can be grouped into four major categories (see Figure 6.1): engine, body, chassis and electrical equipments.

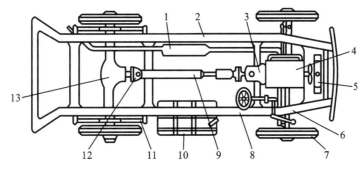

Figure 6.1　Layout of a car

1—exhaust/silencer engine; 2—frame; 3—clutch and gearbox; 4—engine;
5—radiator; 6—front axle; 7—wheel; 8—steering; 9—propeller shaft; 10—fuel tank;
11—road springs; 12—universal joint; 13—differential rear axle

1. Engine

In all automobile components, an automobile engine is the most complicated assembly with dominant effects on the function of an automobile. So, the engine is generally called the "heart" of an automobile. The internal combustion engine is most common; it obtains its power by burning a liquid fuel inside the engine cylinder. There are two types of internal combustion engine: gasoline engine (also called a spark-ignition engine) and diesel engine (also called a compression-ignition engine). The burning fuel generates heat which causes the gas inside the cylinder to increase its pressure and supply power to rotate a shaft connected to the transmission.

All engines have fuel, exhaust, cooling, and lubrication systems. Gasoline engines also have an ignition system. The ignition system supplies the electric spark needed to ignite the air-fuel mixture in the cylinders. When the ignition switch is turned on, current flows from the 12V storage battery to the ignition coil. The coil boosts the voltage to 20,000V, thus the strong spark is produced which ignite the engine fuel.

The fuel system stores liquid fuel and delivers it to the engine. The fuel is stored in the tank, which is connected to a fuel pump by a fuel line. The fuel is pumped from the fuel tank through the fuel lines. It is forced through a filter into the carburetor where it is mixed with air to form a combustible mixture.

Engine cooling must maintain a stable operating temperature, not too hot and not too cold. With the massive amount of heat that is generated from the combustion process, if the engine did not have a method for cooling itself, it would quickly self-destruct. Major engine parts can warp causing oil and water leaks and the oil will boil and become useless.

While some engines are air-cooled, the vast majority of engines are liquid cooled. The water pump circulates coolant throughout the engine, hitting the hot areas around the cylinders and then sends the hot coolant to the radiator to be cooled off.

The lubrication system is important in keeping the engine running smoothly. Motor oil is the lubricant used in the system. The main functions of lubricant system can not only cut down friction by coating moving parts with oil, but also produces a seal between the piston rings and the cylinder walls. It can also carry away sludge and cool the engine by circulating the motor oil. To keep this system working efficiently, oil filters and motor oil must be changed regularly. All other moving parts in an automobile must also be lubricated.

2. Body

An automobile body is a sheet metal shell with windows, doors, a hood, and a trunk deck built into it. It provides a protective covering for the engine, passengers, the cargo. The body is designed to keep the passengers safe and comfortable. The body styling provides an attractive, colorful, modern appearance for the vehicle. It is streamlined to lessen wind resistance and to keep the car from swaying at the driving speeds.

3. Chassis

The chassis is an assembly of those systems that are the major operating parts of a vehicle. The chassis includes the power train, suspension, steering, and brake systems.

The power train system comprises clutch, transmission, propeller shaft, rear axle and differential and the driving road wheels.

The clutch is a friction device used to connect and disconnect a driving force from a driven member. This action may be manual or automatic.

The main purpose of the transmission or gearbox is to provide a selection of gear ratios between the engine and driving conditions. Gear selection may be done manually by the driver or automatically by a hydraulic control system.

The function of the propeller (drive) shaft is to transmit the driver's operation from the gearbox to the input shaft of the rear axle and differential assembly. Flexible joints allow the rear axle and wheels to move up and down without affecting operation.

The rear axle and differential unit transmits the engine's rotational power through 90° from propeller shaft to axle shaft to road wheels. A further function is to allow each driving wheel to turn at different speed, it is essential when cornering because the outer wheel must turn further than inside wheel. A third function is to introduce another gear ratio for torque multiplication.

The basic job of the suspension system is to absorb the shocks caused by irregular road surfaces that would otherwise be transmitted to the vehicle and its occupants, thus helping to keep the vehicle on a controlled and level course, regardless of road conditions.

The steering system, under the control of the driver at the steering wheel, provides the means by which the front wheels are directionally turned. The steering system may be power assisted to reduce the effort required to turn the steering wheel and make the vehicle easier to manoeuvre.

The braking system on a vehicle has three main functions. It must be able to reduce the speed of the vehicle, when necessary; it must be able to stop the car in as short a distance as possible; it must be able to hold the vehicle stationary. The braking action is achieved as the result of friction developed by forcing a stationary surface (the brake lining) into contact with a rotating surface (the drum or disc). Each wheel has a brake assembly, of either the drum type or the disc type, hydraulically operated when the driver applies the foot brake pedal.

4. Electrical equipment

The electrical system supplies electricity for the ignition, horn, lights, heater, and starter. The electricity level is maintained by a charging circuit. This circuit consists of a battery, and an alternator (or generator). The battery stores electricity. The alternator changes the engine's mechanical energy into electrical energy and recharges the battery.

New Words and Expressions

1. component　成分,组成,部件,零件
2. chassis　底盘
3. transmission　变速器,传动,传动系统,传送
4. lubrication　润滑
5. ignition　点火,点燃;点火装置
6. carburetor　化油器
7. combustible　易燃的
8. sludge　软泥,淤泥
9. hood　发动机罩
10. capacity　容量,生产量,才能,能力
11. suspension　悬架
12. differential　差速器
13. clutch　离合器
14. hydraulic　液压的,水压的,水力的
15. pedal　踏板
16. spark-ignition engine　点燃式发动机
17. compression-ignition engine
　　压燃式发动机
18. storage battery　蓄电池
19. ignition coil　点火线圈
20. charging circuit　充电电路
21. motor oil　机油
22. sheet metal　钢板,金属板

Notes

(1) The main functions of lubricant system can not only cut down friction by coating moving parts with oil, but also produces a seal between the piston rings and the cylinder walls.

润滑系统不但可以通过在运动零件上涂油而减小摩擦,还可以通过在活塞环和汽缸壁之间储存机油,产生密封的作用。

(2) The basic job of the suspension system is to absorb the shocks caused by irregular road surfaces that would otherwise be transmitted to the vehicle and its occupants, thus helping to keep the vehicle on a controlled and level course, regardless of road conditions.

悬挂系统的基本功能是吸收不平坦路面引起的震动,否则该震动会被传递给车辆和乘坐人员。因此,悬挂系有助于在各种路况下保持车辆可控并水平行驶。

Translation Skills

被动语态的译法

英语有别于汉语的特点之一,就是被动语态的应用广泛。这一特点在科技英语中反映更为突出。因为在被动语态下,所论证、说明的科技问题在句子的主语位置上,这样就能引起人们的注意。此外,被动语态比主动语态的主观色彩更少,这正是科技作品所需要的。因此,在科技英语中,凡是不需要或不能指出行为主体的场合或在需要突出行为客体的场合,都使用被动语态。但汉语的被动语态使用的范围要窄得多,因为汉语具有英语所没有的无主句,许多被动语态可以采用无主句来表达,而且译法灵活多样。被动语句常见的译法有以下几种。

（一）在谓语前加"被""受""由"字

［例1］ Energy from different sources has been used to do useful work.

译文:各种能源都用来做有用功。

［例2］ When a spring is tightly stretched, it is ready to do work.

译文:当弹簧被拉紧时,它随时可以做功。

［例3］ Once the impurities have been removed, the actual reduction to the metal is an easy step.

译文:杂质一旦被除掉,金属的真正还原就容易了。

［例4］ Beside voltage, resistance and capacitance, alternating current is also influenced by inductance.

译文:除了电压、电阻和电容以外,交流电还受电感的影响。

［例5］ The first stage of a rocket is thrown away only minutes after the rocket takes off.

译文:火箭的第一级在火箭起飞后几分钟就被扔掉了。

（二）在谓语前省略"被"字

在不出现行为主语的被动句中,当被动意义很明显时,汉语习惯不用"被"字,例如,汉语说"工件加工完了",而不说"工件被加工完了"。这种不用"被"字的被动句比用"被"字的被动句要多得多。

[例1] Plastics have been applied to mechanical engineering.

译文:塑料已应用于机械工程领域。

[例2] All radio sets used large, heavy vacuum tubes before transistors were invented.

译文:在晶体管发明之前,所有收音机都使用大而重的真空管。

[例3] Much progress has been made in electrical engineering in less than a century.

译文:不到一个世纪,电气工程就取得了很大的进展。

[例4] Since numerical control was adopted at machine tools, the productivity has been raised greatly.

译文:自从机床采用数控技术以来,生产率大大提高了。

[例5] The head stock is mounted at the left end of the lathe bed.

译文:主轴箱安装在车床床身的左端。

(三) 在行为主体前加"被""由""受""为……所"等字

当句中出现行为主体时,可以在其前面加"被"字来表示被动意义。

[例1] Steam, oil, and water power have been used by mankind for doing work.

译文:蒸汽、石油、水力已被人们用来做功。

[例2] The magnetic filed is produced by a electric current.

译文:磁场由电流产生。

[例3] Magnetic substances are those that are attracted by magnets.

译文:磁性物质就是受磁铁吸引的物质。

[例4] Only a small part of the sun's energy reaching the earth is used by us.

译文:传到地球的太阳能只有一小部分为我们所利用。

[例5] Air is attracted by the earth as every other substance.

译文:空气像任何其他物质一样被地球吸引着。

(四) 译成"是……的"结构

汉语中还有一种不用"被"字的被动句,这就是具有"是……的"结构的句子。凡是着重说明一件事情是"怎样做的"或"什么时候、什么地点做的"等时,就不用"被"字,而用"是……的"结构。

[例1] Iron is extracted from the ore of the blast furnace.

译文:铁是用高炉从铁矿中提炼出来的。

[例2] The material first used was copper because it was easily obtained in its pure state.

译文:最先使用的材料是铜,因为纯铜易于获得。

[例3] Although electricity was discovered about 2000 years ago, it came into practice long after.

译文:虽然电是在约两千年前发现的,但是之后很久以后它才得到实际应用。

[例4] The kind of device is much needed in the mechanical watch making industry.

译文:这种装置在机械表制造业中是很需要的。

(五) 译成汉语主动句

1. 更换主语

翻译时把原文的主语变成宾语,而把行为的主体或相当于行为主体的介词宾语译成主语。

［例1］ More than one hundred elements have been found by chemical workers at present.

译文：目前化学工作者已经发现了一百多种化学元素。

［例2］ A right kind offuel is needed for in atomic reactor.

译文：原子反应堆需要一种适宜的燃料。

［例3］ Two factors,force and distance,are included in the units of work.

译文：功的单位包括两个因素，即力和距离。

［例4］ Semiconductors have been known to mankind for many years.

译文：人们知道半导体已经有很多年了。

［例5］ In chemistry,symbols are used to represent elements.

译文：在化学中用符号代表元素。

2．增加主语

翻译时，把原文的主语译成宾语，并增译泛指性的主语，如"大家""人们""我们""有人"等。

［例1］ All bodies on the earth are known to possess weight.

译文：大家都知道地球上的一切物质都有重量。

［例2］ To explore the moon's surface,rockets were launched again and again.

译文：为了探测月球的表面，人们一次又一次地发射火箭。

［例3］ When water falls a great distance,energy is known to change from potential to kinetic.

译文：我们知道，当水从很高处落下时，能量由位能转变为动能。

［例4］ At one time it was thought that all atoms of the same element were exactly alike.

译文：人们曾经认为同一种元素的所有原子都是完全相同的。

［例5］ It was also said earlier that NC is the operation of machine tools by numbers.

译文：大家早就听说，数控就是机床采用数字操作。

［例6］ With the development of modern electrical engineering, power can be transmitted to wherever it is needed.

译文：随着现代化电气工程的发展，人们能把电能输送到所需要的任何地方。

（六）译成无主句

当无法知道或无须说出行为主体时，往往可以把英语的被动语态译成汉语的无主句。翻译时，把原文的主语译成动词的宾语，有时还可以把原文的主语并入谓语一起翻译。

1．主语译成宾语

［例1］ Work is down,when an object is lifted.

译文：当举起一个物体时，就做了功。

［例2］ To get all the stages off the ground,a first big push is needed.

译文：为了使火箭各级全部离开地面，需要有一个巨大的第一次推力。

［例3］ If 250 million hydrogen atoms were placed side by side,they would extend about one inch.

译文：如果将两亿五千万个氢原子一个接一个排列起来，也不过一英寸（1 in＝25.4 mm）左右长。

［例 4］ So attention is now being given to possible uses of atomic energy.
译文：因此，现在正把注意力放在原子能的可能的用处上。
［例 5］ Mercury freezes if it is cooled to too low a temperature.
译文：如果将水银冷却到过低的温度，就会使它凝固。

2. 主语与谓语合译

英语中有些成语动词含有名词（如 make use of, pay attention to 等），变成被动语态时将该名词译作主语，这是一种特殊的被动语态。翻译时可以把主语和谓语合译，译成汉语无主句的谓语。

［例 1］ Use can be made of laser beams to burn a hole in a diamond.
译文：可以利用激光在金刚石上打孔。
［例 2］ Attention must be paid to safety in handling radio active materials.
译文：处理放射性材料时必须注意安全。
［例 3］ Brief reference should be made here to one of the alkyl oxides.
译文：这里应该简单提一提一种烷基氧化物。
［例 4］ Allowance must, no doubt, be made for the astonishing rapidity of communication in these days.
译文：毫无疑问要考虑到现代通信的惊人速度。
［例 5］ Account should be taken of the low melting point of this substance.
译文：应该考虑到这种物质低熔点。

❖ Reading Material

The History of the Automobile

The history of the automobile begins as early as 1769, with the creation of steam-powered automobiles capable of human transport. Steam-powered self-propelled vehicles are thought to have been devised in the late 18th century. Nicolas-Joseph Cugnot demonstrated an experimental steam-driven artillery tractor, in 1770 and 1771.

By 1784, William Murdoch had built a working model of a steam carriage in Redruth, and in 1801 Richard Trevithick was running a full-sized vehicle on the road in Camborne. Such vehicles were in vogue for a time, and over the next decades such innovations as hand brakes, multi-speed transmissions, and better steering developed.

At the 1900s some steam-powered self-propelled vehicles were commercially successful in providing mass transit, until a backlash against these large speedy vehicles resulted in passing a law, the Locomotive Act, in 1865 requiring self-propelled vehicles on public roads in the United Kingdom be preceded by a man on foot waving a red flag and blowing a horn. The law was not repealed until 1896, although the need for the red flag was removed in 1878.

German engineer Karl Benz, inventor of numerous car-related technologies, is generally regarded as the inventor of the modern automobile. The four-stroke petrol (gasoline) internal combustion engine that constitutes the most prevalent form of modern automotive propulsion is a creation of German inventor Nikolaus Otto.

Early attempts at making and using internal combustion engines were hampered by the lack of suitable fuels, particularly liquids, and the earliest engines used gas mixtures. A later version was propelled by coal gas. Karl Benz built his first automobile in 1885 in Mannheim. Benz was granted a patent for his automobile on 29 January 1886, and began the first production of automobiles in 1888. Soon after, Gottlieb Daimler and Wilhelm Maybach in Stuttgart in 1889 designed a vehicle from scratch to be an automobile, rather than a horse-drawn carriage fitted with an engine.

1. Veteran car era

By 1900, mass production of automobiles had begun in France and the United States. By the start of the 20th century, the automobile industry was beginning to take off in western Europe, especially in France, they produced 30,204 in 1903, representing 48.8% of world automobile production that year.

Innovation was rapid and rampant, with no clear standards for basic vehicle architectures, body styles, construction materials, or controls. Automobiles were seen as more of a novelty than a genuinely useful device. Breakdowns were frequent, fuel was difficult to obtain, roads suitable for traveling were scarce, and rapid innovation meant that a year-old car was nearly worthless.

Major breakthroughs in proving the usefulness of the automobile came with the historic long-distance drive of Bertha Benz in 1888, when she traveled more than 80 kilometers (50 mile) from Mannheim to Pforzheim, to make people aware of the potential of the vehicles her husband, Karl Benz, manufactured.

2. Edwardian era

Edwardian era lasted from roughly 1905 through to the beginning of World War Ⅰ in 1914. Throughout this era, development of automotive technology was rapid, due in part to a huge number (hundreds) of small manufacturers all competing to gain the world's attention.

Key developments included electric ignition system (by Robert Bosch, 1903), independent suspension, and four-wheel brakes (by the Arrol-Johnston Company of Scotland in 1909). Transmissions and throttle controls were widely adopted, allowing a variety of cruising speeds, though vehicles generally still had discrete speed settings, rather than the infinitely variable system familiar in cars of later eras.

3. Vintage era

The vintage era lasted from the end of World War Ⅰ (1919), through the Wall Street Crash at the end of 1929. During this period, the car came to dominate, with closed bodies and standardized controls becoming the norm. In 1919, 90% of cars sold were open; by 1929, 90% were closed.

Development of the internal combustion engine continued at a rapid pace, with multi-valve and overhead camshaft engines produced at the high end, and V8, V12, and even V16 engines conceived for the ultra-rich.

4. Pre-WW Ⅱ era

The pre-war era began with the Great Depression in 1930, and ended with the recovery after World War Ⅱ, commonly placed at 1948. By the 1930s, most of the mechanical technology used in today's automobiles had been invented. After 1930, the number of auto manufacturers declined sharply as the industry consolidated and matured.

5. Post-war era

Automobile design finally emerged from the shadow of World War Ⅱ in 1949. The strut-suspended 1951 Ford Consul joined the 1948 Morris Minor and 1949 Rover P4 in waking up the automobile market in the United Kingdom. In Italy, Enzo Ferrari was beginning his 250 series.

Throughout the 1950s, engine power and vehicle speeds rose, and cars spread across the world. The market changed somewhat in the 1960s, as Detroit began to worry about foreign competition, the European makers adopted ever-higher technology, and Japan appeared as a serious car-producing nation. BMC's revolutionary space-saving Mini, which first appeared in 1959, captured large sales world-wide.

In America, performance became a prime focus of marketing, exemplified by pony cars and muscle cars. In 1964 the popular Ford Mustang appeared. But everything changed in the 1970s as the 1973 oil crisis, automobile emissions control rules, Japanese and European imports. Small performance cars from BMW, Toyota, and Nissan took the place of big cars from America and Italy.

The biggest developments of the era were the widespread use of independent suspensions, wider application of fuel injection, and an increasing focus on safety in the design of automobiles.

6. Modern era

The modern era is normally defined as the 25 years preceding the current year. Without considering the future of the car, the modern era has been one of increasing standard, platform sharing, and computer-aided design.

Some particularly notable advances in modern times are the widespread of front-wheel drive and all-wheel drive, the adoption of the V6 engine configuration, and the ubiquity of fuel injection. Body styles have changed as well in the modern era. Three types, the hatchback, minivan, and sport utility vehicle, dominate today's market, yet are relatively recent concepts.

The rise of pickup trucks in the United States, and SUVs worldwide has changed the face of motoring, with these "trucks" coming to command more than half of the world automobile market. The modern era has also seen rapidly rising fuel efficiency and engine output.

New Words and Expressions

1. hand brake   手刹
2. internal combustion engine   内燃机
3. patent   授予专利,专利的,专利权
4. electric ignition system   电火花点火系统
5. independent suspension   独立悬挂系统

Lesson 7 Manufacturing Process

Text

Manufacturing is to utilize and manage materials, people, equipment, and money to produce products. To achieve a successful and economical manufacturing, the planning must be made from the beginning of the design stage and go through the selection of materials, processes, equipment, and the scheduling of production. These processes constitute the so-called manufacturing process. The process of a product involves: product design and prepare; transport of the raw and processed materials; roughcast forming; machining; deburring; heat treatment; assembly; test; packing and painting, which is usually completed by the coordinated work of different factories or workshops.

The process of a product from raw materials to finished parts:

• Forming of the roughcast or part—casting, forging, punching, welding, pressing, agglomeration, infusing and molding, and so on.

• Machining—cutting, grinding, special machining (non-conventional machining).

• Material's properties adaption and treatment—heat treatment, electroplate, coating and filming.

• Assembling—connecting the parts according to the definite relationship and specification, including part fixing, connection, adjustment, balance, inspection and test.

The purpose of machining is to turn a kind of raw material into the finished parts that meets the requirement of the product. Under given manufacturing conditions, how to apply both an economical and efficient method, as well as a rational process routine, to achieve the desired parts is the concern of manufacturing process.

The five basic techniques of machining metal include drilling, boring, turning, planning, milling and grinding. Variations of the five basic techniques are employed to meet special situations (see Table 7.1).

Drilling consists of cutting a round hole by means of a rotating drill. Boring, on the other hand, involves the finishing of a hole already drilled or cored by means of a rotating, offset, single-point tool. On some boring machines, the tool is stationary and the work revolves, on others, the reverse is true.

Table 7.1 Comparison of basic machining operations for ductile materials

Operation	Shape produced	Machine tool	Cutting tool	Relative motion Tool	Relative motion Work	Surface roughness /μm	Min. Prod. Tolerance/in
Turning (external)	Surface of revolution (cylindrical)	Lath, boring machine	Single point	↔ ↕		32~500	±0.001
Boring (internal)	Cylindrical (enlarge hole)	Boring machine	Single point	↔		16~250	±0.0001 ±0.001
Shaping and planning	Flat surfaces or slots	Shaper, planer	Single point	↕ ↔	↔	32~500	±0.001
Drilling	Cylindrical (originates holes 0.010 to 4 in. Dia.)	Drill press	Drill: twin edges	↕	fixed.	125~250	±0.002
Milling end, form face, slab	Flat and contour surface and slots	Milling machine	Multiple points (cutter teeth)			32~500	±0.001
Grinding cylindrical surface plunge	Cylindrical and flat	Grinding machine	Multiple points (grind wheel)			8~125	±0.0001

The lathe, as the turning machine is commonly called, is the father of all machine tools. The piece of metal to be machined is rotated and the cutting tool is advanced against it.

Planning metal with a machine tool is a process similar to planning wood with hand plane. The essential difference lies in the fact that the cutting tool remains in a fixed position while the work is moved back and forth beneath it. Planers are usually large piece of equipment; sometimes large enough to handle the machining of surfaces 15 to 20 feet wide and twice as long. A shaper differs from a planer in that the work-piece is held stationary and the cutting tool travels back and forth.

After lathes, milling machines are the most widely used for manufacturing applications. Milling consists of machining a piece of metal by bringing it into contact with a rotating cutting tool which has multiple cutting-edges. There are many types of milling types of milling machines designed for various kinds of work. Some of the shapes produced by milling machines are extremely simple, like the slots and flat surfaces produced by circular saws. Other shapes are more complex and may consist of a variety of combinations of flat and curved surfaces depending on the shape given to the cutting-edges of the tool and on the travel path of the tool.

Grinding consists of shaping a piece of work by bringing it into contact with a rotating abrasive wheel. The process is often used for the final finishing to close dimensions of a part that has been heat-treated to make it very hard.

New Words and Expressions

1. manufacturing process   制造工艺
2. deburring   修边,去毛刺
3. casting   铸造
4. forging   锻造
5. punching   冲压
6. welding   焊接
7. pressing   挤压
8. agglomeration   烧结
9. infusing   注塑
10. molding   制模,塑造
11. electroplate   电镀
12. drilling   钻削
13. boring   镗削
14. turning   车削
15. planning   刨削
16. milling   铣削
17. grinding   磨削

Notes

(1) The process of a product involves: product design and prepare; transport of the raw and processed materials; roughcast forming; machining; deburring; heat treatment; assembly; test; packing and painting, which is usually completed by the coordinated work of different factories or workshops.

产品的制造过程包括:产品的设计和准备、原材料和已加工材料的运输、毛坯的成形、切削加工、去毛刺、热处理、装配、检验、包装和喷漆等,这些通常是由不同企业或者车间共同完成的。

(2) Planning metal with a machine tool is a process similar to planning wood with hand plane. The essential difference lies in the fact that the cutting tool remains in a fixed position while the work is moved back and forth beneath it. Planers are usually large piece of equipment; sometimes large enough to handle the machining of surfaces 15 to 20 feet wide and twice as long. A shaper differs from a planer in that the work-piece is held stationary and the cutting tool travels back and forth.

用龙门刨床刨削金属的过程类似于用手刨刨木头。二者的本质区别实际上在于,用龙门刨床刨削时,刨刀在一个位置固定不动,而工件在刨刀下面往复移动。龙门刨床通常是大型设备,有时大到是以加工宽达15~20 ft,而长是宽的两倍的表面。牛头刨床与龙门刨床的区别在于它的工件是固定的,而刀具往复移动。

❖ Translation Skills

句量的增减——分译与合译

一、分译法

由于汉语习惯上一般不使用长句,因此在英译汉时,往往要把原文中较长的句子成分或不易安排的句子成分拆开来,译成汉语的分句或从句,以使译文通顺流畅,层次分明,符合汉语的表达习惯。

(一) 单词的分译

有的英文句子虽不长,但照译时译文很别扭,读起来非常拗口。尤其是原文中个别单词很难处理,不拆开翻译就不通顺或容易发生误解,这时可把这种单词译成一个分句或单句。

1. 副词的分译

某些表示主观评价的副词往往必须分译。

[例1] The time could have been *more profitably* spent in making a detailed investigation.

译文:如果当初把时间花在细致的调查研究上,益处就更大了。

[例2] This year the estimated production of transistors is 400 million, and we can *safely* assume that even this tremendous number is likely to increase.

译文:今年晶体管的预期产量是4亿只。这个数字虽然已经很大,但产量可能还要增加。我们这样估计是有把握的。

2. 形容词的分译

[例1] Alchemists made *resultless* efforts to transform one metal into another.

译文:炼金术士企图把一种金属变成另一种金属,这完全是徒劳的。

[例2] He spoke with *understandable* pride of the invention of the instrument.

译文:他自豪地谈到那种仪器的发明,这种自豪感是可以理解的。

3. 名词的分译

[例1] The *price* limits its production.

译文:它价钱昂贵,这限制了它的生产量。

[例2] The *dust*, the *uproar* and the growing dark threw everything into chaos.

译文:滚滚的烟尘,嘈杂的人声,渐浓的夜色,使一切都陷入混乱之中。

(二) 短语的分译

短语的分译是指把原文中一个短语译成句子,从而将原文的一个句子分译成两个或两个以上的句子。这种现象在英译汉中比单词的分译更为常见。英语中能提出来分译的短语通常有分词短语、介词短语、动词不定式短语和名词短语。

1. 分词短语(包括分词独立结构)的分译

[例1] Let us have a good look at Figure 2 *showing how heat makes objects expand*.

译文:我们仔细看一下图2,它说明了热是如何使物体膨胀的。

[例2] *Science and technology modernized*, industry and agriculture will develop rapidly.

译文：实现科学技术的现代化，工、农业就会迅速发展。

2. 介词短语的分译

［例1］ Copper is an important conductor, both *because of its high conductivity* and *because of its abundance and low cost*.

译文：铜是一种重要导体，因为它的导电率高，资源丰富，价格又低。

［例2］ The factory was already spreading a fame *for its products*.

译文：这个工厂因为产品质量优良，已远近闻名。

3. 动词不定式短语的分译

［例1］ Einstein's theory of relatively is too difficult *for the average mind to understand*.

译文：爱因斯坦的相对论太难，一般人无法理解。

动词不定式短语作宾语补足语或主语补足语时，如果谓语动词是 know, find, consider 等，谓语动词可以分译，动词不定式短语可与其宾语或主语一起译成从句。

［例2］ The weight of an object has been found *to vary from place to place on the earth*.

译文：人们发现，物体的重量在地球上因地而异。

4. 名词短语的分译

［例1］ *The wrong power-line connections* will damage the motor.

译文：如果把电源接错，就会损坏电动机。

［例2］ But *the exigency of the case* admitted of no alternatives.

译文：但情况紧急，别无选择。

（三）句子的分译

所谓句子的分译，指的是把原文的从句译成分句，或把一个简单句拆开，译成两个或两个以上的句子，以及在翻译复句过程中增加分句或句子的数量。

1. 定语从句的分译

把定语从句与主句分开，译成后置的并列分句或独立句，这是定语从句汉译时的一种重要方法。

［例1］ The control unit is an important part of the computer, *which can cause the machine to operate according to man's wishes*.

译文：控制装置是计算机的重要部件，它能使机器按照人的意志进行操作。

［例2］ In the future, robots *which can do almost all of our hard work* will be seen working everywhere.

译文：在将来，到处都会看到机器人在工作，它们能替我们担负起几乎所有的繁重工作。

2. 单句的分译

单句的分译就是各种句子成分的分译，即把一个句子译成两个或两个以上的句子。

［例1］ Carbon acid is particularly familiar to us in the form of soda water.

译文：汽水中就有碳酸。我们对碳酸的这种形式很熟悉。

［例2］ Two of the advantages of the transistor are its being small in size and its being able to be put close to each other without overheating.

译文：晶体管有两个优点，一是体积小，二是能靠近放置而不过热。

3. 复句的分译

复句的分译是指在翻译复句的过程中增加分句或句子的数量,即得原文的复句中有的分句译成两个或两个以上的分句。

[例] The light speed being known, it would be possible to decide the distance from the earth to the moon provided that a single light beam could be sent to the moon and made to return to the earth's receiving station.

译文：因为光速是已知的,所以只要能将一束光送上月球并使它返回地球接收站,就有可能测定从地球到月球的距离。

二、合译法

(一) 单句的合译

单句的合译往往以一个单句的翻译为基础,把其他单句的内容压缩成一两个句子成分加到译文中去。

[例] The chemical makes frequent use of catalysts. He adds them in small quantities to reacting materials.

译文：化学工作者经常通过把少量催化剂加入反应物中来利用催化剂。

(二) 主从复合句的合译

对于有些主从复合句,在翻译时可将从句译成词组,因而译文就成了简单句。

[例1] By then ironworkers had not learned how to heat the iron ore enough *that they could melt it*.

译文：那时候,炼铁工人还不知道如何把铁矿石加热而使其熔化。

[例2] Materials to be used for structural purposes are chosen *so that they behave elastically in the environmental conditions*.

译文：结构材料的选择应使其在外界条件下保持弹性。

[例3] *It is necessary that a qualified engineer be given a good training in general science*, particularly in mathematics, physics and chemistry.

译文：一个合格的工程师必须接受普通科学,尤其是数学、物理和化学方面的良好教育。

(三) 并列复合句的合译

[例1] It was in mid-August, and the repair section operated under the blazing sun.

译文：八月中旬,修理组人员在烈日下工作。

[例2] The Post Office was helpful, and Marconi applied in June 1896 for the world's first radio patent.

译文：在英国邮局的帮助下,马克尼于1896年6月申请了世界上第一项无线电专利。

[例3] Sulfuric acid enters into the manufacture of explosives, dyestuffs and drugs; it is used in sugar refining and in the preparation of fertilizers.

译文：硫酸被用来制造炸药、燃料、药物,以及制作糖和化肥等。

❖ Reading Material

Turning Methods and Machines

1. Introduction

The basic engine lathe, which is one of the most widely used machine tools, is very versatile when used by a skilled machinist. However, it is not particularly efficient when many identical parts must be machined as rapidly as possible.

The turret lathe and automatic lathe in their various forms have been developed and improved with the objectives of producing machined parts more rapidly and accurately at lower cost. Mechanical power, in electrical, hydraulic, or pneumatic form, has replaced human muscle power for such functions as feeding tools, operating chucks or collets, and feeding bar stock in the machine.

2. Lathes and lathe components

Of the many standards and special types of turning machines, the most important, most versatile, and most widely recognized is the engine lathe. The modern engine lathe provides a wide range of speeds and feeds which allow optimum settings for almost any operation.

The components of an engine lathe is shown in Figure 7.1.

Headstock The headstock is the powered end and is always at the operator's left. It contains the speed changing gears and the revolving, driving spindle, to which any one of several types of work holders is attached. The center of the spindle is hollow so that long bars may be put through it for machining.

Tailstock The tailstock on the hardened ways is non-rotating. It can be moved left or right to adjust to the length of the work.

Carriage The carriage can be moved left or right either by hand wheel or power feed. This provides the motion along the z-axis. During this travel turning cuts are made.

Cross slide The cross slide is mounted on the carriage and can be moved in and out (x-axis) perpendicular to the carriage motion.

Compound rest The compound rest is mounted on the carriage. It can be moved in and out by its hand wheel for facing or for setting the depth of cut.

Tool-post The tool post is mounted on the compound rest. It can also be rotated so as to present the cutter to the work at whatever angle is best for the job.

Bed The bed of the lathe is its "backbone". It must be rigid enough to resist deflection in any direction under the loads. The bed is made of cast iron or a steel weldment, in a box or I-beam shape.

Ways The ways of the lathe are the flat or V-shaped surfaces on which the carriage and the tailstock are moved left and right.

Lesson 7　Manufacturing Process

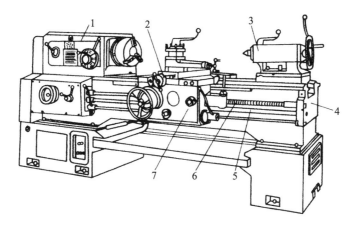

Figure 7.1　The engine lathe

1—headstock; 2—tool-post; 3—tailstock; 4—bed; 5—feed-shaft; 6—lead-screw; 7—carriage

3. Turret lathe

The standard engine lathe is versatile, but it is not a high production machine. When production requirements are high, more automated turning machines must be used.

The turret lathe represents the first step from the engine lathe toward the high production turning machines. The turret lathe is similar to the engine lathe except that tool-holding turrets replace the tailstock and the tool post compound assembly. The difference between the engine and turret lathes is that the turret lathe is adapted to quantity production work, whereas the engine lathe is used primarily for miscellaneous jobbing or single-operation work.

4. Cutting motions and parameters

For example, in turning, the work-piece rotates, the cutting tool moves longitudinally, the mixture of the two motions forms the external cylindrical surface. There are three surface created in turn during the new surface is forming (see Figure 7.2). There are three surfaces and two motions. Surface to be machined means the surface in which the material is to be removed; machining surface means the surface is being machined by the cutting edge; machined surface means the surface in which the material has been removed.

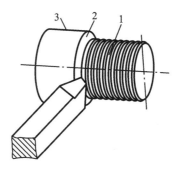

Figure 7.2　Cutting motions in turning

1—machined surface; 2—machining surface; 3—surface to be machined

5. Cutting motion

The basic motion for a machine tool includes linear motion and rotational motion. If classified according to the functions of the cutting tool in relation with the workpiece, they are called the main motion and the feed motions.

The main motion required to remove the metal. Usually it has the biggest velocity, and consumes most of the power.

The feed motion brings the new metal into cutting constantly. It can be continuous or interrupted.

6. Lathe cutting tools

The shape and geometry of the lathe tools depend upon the purpose for which they are employed. Turning tools can be classified into two main groups, namely, external cutting tools and internal cutting tools. Each of theses two groups include the following types of tools:

Turning tools　Turning tools can be either finishing or rough turning tools. Rough turning tools have small nose radii and are employed when deep cuts are made. On the other hand, finishing tools have larger nose radii and are used for obtaining the final required dimensions with good surface finish by making slight depth of cut. Rough turning tools can be right-hand or left-hand types, depending upon the direction of feed. They have straight, bent, of offset shanks.

Facing tools　Facing tools are employed in facing operations for machining plane side or end surface. These are tools for machining left-hand-side surfaces and tools for right-hand-side surfaces. Those side surfaces are generated through the use of the cross feed, contrary to turning operations, where the usual longitudinal feed is used.

Cutoff tools　Cutoff tools, which are sometimes called parting tools, serve to separate the workpiece into parts and/or machine external annual grooves.

Thread-cutting tools　Thread-cutting tools have either triangular, square, or trapezoidal cutting edges, depending upon the cross section of the desired thread. Also, the plane angles of these tools must always be identical to those of the thread forms. Thread-cutting tools have straight shanks for external thread cutting and are of the bent-shank type when cutting internal threads.

Form tools　Form tools have edges especially manufactured to take a certain form, which is opposite to the desired shape of the machined workpiece. An HSS tool is usually made in the form of a single piece, contrary to cemented carbides or ceramic, which are made in the form of tips. The latter are brazed or mechanically fastened to steel shanks.

New Words and Expressions

1. turret lathe　转塔车床
2. headstock　主轴箱
3. tailstock　尾座
4. carriage　溜板箱

5. cross slide   横向托板
6. compound rest   复式刀架
7. tool post   刀架
8. bed   床身
9. way   导轨
10. cutting motion   切削运动
11. radii   半径
12. facing tool   端面车刀
13. cutoff tool   切断车刀
14. thread-cutting tool   螺纹车刀
15. form tool   成形刀

Lesson 8 Tolerance and Interchangeability

Text

Quality and accuracy are major considerations in making machine parts or structures. Interchangeable parts require a high degree of accuracy to fit together. With increasing accuracy or less variation in the dimension, the labor and machinery required to manufacture a part is more cost intensive. Any manufacturer should have a thorough knowledge of the tolerances to increase the quality and reliability of a manufactured part with the least expense.

An engineering drawing must be properly dimensioned in order to convey the designer's intent to the end user. Dimensions of parts given on blueprints and manufactured to those dimensions should be exactly alike and fit properly. Unfortunately, it is impossible to make things to an exact dimension. Most dimensions have a varying degree of accuracy and a means of specifying acceptable limitations in dimensional variance so that a manufactured part will be accepted and still function. It is necessary that the dimensions, shapes and mutual position of surfaces of individual parts are kept within a certain accuracy to achieve their correct and reliable function. Routine production processes do not allow maintenance (or measurement) of the given geometrical properties with absolute accuracy. Actual surfaces of the produced parts therefore differ from ideal surfaces prescribed in drawings. Deviations of actual surfaces are divided into four groups to enable assessment, prescription and checking of the permitted inaccuracy during production:

- Dimensional deviations;
- Shape deviations;
- Position deviations;
- Surface roughness deviations.

As mentioned above, it is principally impossible to produce machine parts with absolute dimensional accuracy. In fact, it is not necessary or useful. It is quite sufficient that the actual dimension of the part is found between two limit dimensions and a permissible deviation is kept with production to ensure correct functioning of engineering products. The required level of accuracy of the given part is then given by the dimensional tolerance which is prescribed in the drawing. The production accuracy is prescribed with regards to the functionality of the product and to the economy of production as well. The principal factor used to set a tolerance for a dimension should be the function of the feature being controlled by the dimension. Unnecessarily tight tolerances lead to high cost of manufacture resulting from more expensive manufacturing methods and higher reject rates.

Lesson 8 Tolerance and Interchangeability

1. Dimension and tolerance

In dimensioning a drawing, the numbers placed in the dimension lines represent dimensions that are only approximate and do not represent any degrees of accuracy unless so stated by the designer. The numbers are termed nominal size. Designers typically specify a component's nominal dimensions such that it fulfills its requirements. In reality, components cannot be made repeatedly to nominal dimensions, due to surface irregularities and the intrinsic surface roughness. Some variability in dimensions must be allowed to ensure manufacture is possible. However, the variability permitted must not be so great that the performance of the assembled parts is impaired. The allowed variability on the individual component dimensions is called the tolerance.

The control of dimensions is necessary in order to ensure assembly and interchangeability of components. Tolerances are specified on critical dimensions that affect the clearance and interference fits. One method of specifying tolerances is to state the nominal dimension followed by the permissible variation. An example of a dimension could be stated as 40.000 ± 0.003 mm, which means that the processing dimension must be between 39.997 mm and 40.003 mm. If the variation can vary either side of the nominal dimension, the tolerance is called a bilateral tolerance. If one tolerance is zero, the tolerance is called a unilateral tolerance.

Most organizations have general tolerances that apply to dimensions when an explicit dimension is not specified on a drawing. For processing dimensions a general tolerance may be ± 0.5 mm, so a dimension specified as 15.0 mm may range between 14.5 mm and 15.5 mm. Other general tolerances can be applied to features such as angles, drilled and punched holes, castings, forgings, weld beads and fillets.

2. Standard fits for holes and shafts

A standard engineering task is to determine tolerances for a cylindrical component, e. g. a shaft, fitting or rotating inside a corresponding cylindrical component or hole. The tightness of fit will depend on the application. For example, a gear located onto a shaft would require a "tight" interference fit, where the diameter of the shaft is actually slightly greater than the inside diameter of the gear hub in order to be able to transmit the desired torque. Alternatively, the diameter of a journal bearing must be greater than the diameter of the shaft to allow rotation. Given that it is not economically possible to manufacture components to exact dimensions, some variability in sizes of both the shaft and hole dimension must be specified. However, the range of variability should not be so large that the operation of the assembly is impaired. Rather than having an infinite variety of tolerance dimensions that could be specified, national and international standards have been produced defining bands of tolerances, examples of which are listed in Table 8.1, e. g. H11/C11. To turn this information into actual dimensions, corresponding tables exist defining the tolerance levels for the size of dimension under consideration. In order to use this information, the following list and Figure 8.1 give definitions used in conventional tolerance. Usually the hole-based system is used, as this results in a reduction in the variety of drill, reamer, broach and gauge tooling required within a company.

Table 8.1 Example of tolerance bands and typical applications

Class	Description	Characteristic	ISO code	Assembly	Application
Clearance	Loose running fit	For wide commercial tolerance	H11/c11	Noticeable clearance	IC engine exhaust value in guide
	Free running fit	Good for large temperature variations, high running speeds or heavy journal pressures	H9/d9	Noticeable clearance	Multiple bearing shafts, hydraulic position in cylinder, removable levers, bearing for rollers
	Close running fit	For running on accurate machines and accurate location at moderate speeds and journal pressures	H8/f7	Clearance	Machine tool main bearings, crankshaft and connecting rod bearings, shaft sleeves, clutch sleeves, guide blocks
	Sliding fit	When parts are not intended to run freely, but must move and turn and locate accurately	H7/g6	Push fit without noticeable clearance	Push on gear wheels and clutches, connecting rod bearings, indicator pistons
	Location clearance fit	Provides snug fit for location of stationary parts, but can be freely assembled	H7/h6	Hand pressure with lubrication	Gears, tailstock sleeves, adjusting rings, loose bushes for piston bolts and pipeline
Transition	Location transition fit	For accurate location (compromise between clearance and interference fit)	H7/k6	Easily tapped with hammer	Pulleys, clutches, gears, flywheels
		For more accurate location	H7/n6	Needs pressure	Motor shaft armatures, toothed collars on wheels
Interference	Locational interference fit	For parts requiring rigidity and alignment with accuracy of location	H7/p6	Needs pressure	Split journal bearings
	Medium drive fit	For ordinary steel parts or shrink fits on light sections	H7/s6	Needs pressure or temperature difference	Clutch hubs, bearings, bushes in blocks, wheels, connecting rods, Bronze collars on grey cast iron hubs

Size: a number expressing in a particular unit the numerical value of a dimension.

Actual size: the size of a part as obtained by measurement.

Limits of size: the maximum and minimum sizes permitted for a feature.

Maximum limit of size: the greater of the two limits of size.
Minimum limit of size: the smaller of the two limits of size.
Basic size: the size by reference to which the limits of size are fixed.
Deviation: the algebraic difference between a size and the corresponding basic size.
Actual deviation: the algebraic difference between the actual size and the corresponding basic size.
Upper deviation: the algebraic difference between the maximum limit of size and the corresponding basic size.
Lower deviation: the algebraic difference between the minimum limit of size and the corresponding basic size.
Tolerance: the difference between the maximum limit of size and the minimum limit of size.
Shaft: the term used by convention to designate all external features of a part.
Hole: the term used by convention to designate all internal features of a part.

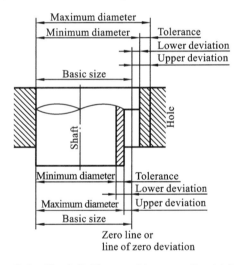

Figure 8.1 The definitions used in conventional tolerance

New Words and Expressions

1. dimension 尺寸,尺度
2. dimensional deviations 尺寸公差
3. shape deviation 形状公差
4. position deviation 位置公差
5. nominal dimension 公称尺寸,名义尺寸
6. surface roughness 表面粗糙度
7. manual 手动的,手工的
8. tolerance 公差
9. shaft 轴
10. assembly 装配
11. clearance fit 间隙配合
12. transition fit 过渡配合
13. interference fit 过盈配合
14. bilateral tolerance 双向(边)公差
15. unilateral tolerance 单向(边)公差

Notes

(1) With increasing accuracy or less variation in the dimension, the labor and machinery required to manufacture a part is more cost intensive.

随着精度的增加和尺寸变动量的减小,制造零件所需的劳动力和机器更趋向成本密集型。

(2) Routine production processes do not allow maintenance (or measurement) of the given geometrical properties with absolute accuracy.

在常规生产过程中不可以对具有绝对精度的给定几何特征进行维修或测量。

(3) In dimensioning a drawing, the numbers placed in the dimension lines represent dimensions that are only approximate and do not represent any degrees of accuracy unless so stated by the designer.

在图纸上标注尺寸时,尺寸线上标注的数字仅仅表示近似的尺寸,而不表示任何精度,除非设计人员加以说明。

❖ Translation Skills

句子成分的转换

在翻译过程中,虽然有时可以使译文的主语、谓语、宾语等句子成分与原文对应,但由于英汉两种语言的表达习惯在大多数情况下并不一一对应,所以不能采取语言对等的译法。为了使译文通顺达意,必须运用转换翻译技巧,包括词性转换、句子成分的转换以及句型转换。

一、主语转换成宾语

原文中的主语在译文中作宾语,并增译主语,这种转换技巧常用于翻译被动句。

[例1] *Salt* is known to have a very strong corroding effect on metals.

译文:大家都知道盐对金属有很强的腐蚀作用。

[例2] Considerable *use* has been made of these data.

译文:这些资料得到了充分的利用。

[例3] *New* computer viruses and logic bombs are discovered every week.

译文:我们每个星期都能发现新型计算机病毒和逻辑炸弹程序。

原文中的主语在译文中作宾语,而原文中的行为主体、用作状语的介词短语或其他词语则相应地译成主语。

[例1] *Computers* are employed *in most libraries*.

译文:大多数图书馆已使用计算机。

[例2] *Two factors*, force and distance, are included in the units of work.

译文:功的单位包括两个因素,即力和距离。

[例3] With the emergence of an electronic currency, *everyone of us* would be affected.

译文:电子货币的出现将影响到我们每个人。

二、主语转换成谓语

英语中,有些动词短语中的名词用于被动语态时在句中作主语,翻译时应把主语和谓语合并,并按原动词短语的意思译出,如 care,need,attention,emphasis,improvement 等名词作主语时,常这样处理。

[例1] *Care* should be taken not to damage the instruments.
译文:注意不要损坏仪器。

[例2] There is a *need* for improvement in our experimental work.
译文:我们的实验工作需要改进。

[例3] *Attention* must be paid to environmental protection in developing economy.
译文:发展经济必须注意环境保护。

三、宾语转换成主语

英语中常用结构如 to have a length of…,to have a height of…,to have a density of…,to have a voltage of…,如果照字面译成汉语,就十分生硬,往往将"长度""高度""密度""电压"等作为主语。

[例1] This sort of stone has *a relative density* of 2.8.
译文:这种石头的相对密度是2.8。

[例2] Light beams can carry more *information* than radio signals because light has a much higher *frequency* than radio waves.
译文:光束携带的信息比无线电信号携带的信息多,因为光波的频率比无线电波高得多。

四、谓语转换成主语

某些英语动词如 act,behave,feature,characterize,relate,load,conduct 等在汉语中往往都要转译成名词才符合汉语的表达习惯,因此,在句子中的成分也就相应地转换成主语。

[例1] A highly developed physical science *is characterized* by an extensive use of mathematics.
译文:一门高度发展的物理学的特点是广泛地应用数学。

[例2] Water with salt *conducts* electricity very well.
译文:盐水的导电性能良好。

五、定语转换成谓语或表语

进行这种转换往往是为了突出定语所表达的内容。

[例1] There is a large amount of energy *wasted* due to friction.
译文:大量的能量由于摩擦而损耗掉了。

[例2] *Many* factors enter into equipment reliability.
译文:涉及设备可靠性的因素很多。

六、定语与状语相互转换

由于被修饰词的词性改变,起修饰作用的定语与状语常常互相转换。此外,修饰全句的状语有时可译成定语。

[例1] We should have a *firm* grasp of the fundamentals of mechanics.
译文:我们应牢固掌握力学的基本知识。

[例2] The electronic computer is *chiefly* characterized by its accurate and rapid computations.
译文:电子计算机的主要特点是计算准确而迅速。

七、其他

主动语态句子的某些状语可转译成主语，there be 句型中修饰主语的 of 短语可从定语转换成主语，to have a length of... 句型中的主语可以转换成定语。

[例1]　Various substances differ widely *in their magnetic characteristics*.

译文：不同物质的磁性大不相同。

[例2]　There are three main laws *of mechanics*, or three laws of Newton.

译文：力学有三大定律，即牛顿三定律。

[例3]　There exist many sources of *energy both potential and kinetic*.

译文：势能和动能都有许多来源。

❖ Reading Material

ISO System of Limits and Fits

The standard ISO is used as an international standard for linear dimension tolerances and has been accepted in most industrially developed countries as a national standard. The ISO system of tolerances and fits can be applied in tolerances and deviations of smooth parts and for fits created by their coupling. It is used particularly for cylindrical parts with round sections. Tolerances and deviations in this standard can also be applied in smooth parts of other sections. Similarly, the system can be used for coupling (fits) of cylindrical parts and fits with parts having two parallel surfaces (e.g. fits of keys in grooves). The term "shaft", used in this standard has a wide meaning and serves for specification of all outer elements of the part, including those elements which do not have cylindrical shapes. Also, the term "hole" can be used for specification of all inner elements regarding of their shape.

In ISO system, each part has a basic size whose limit dimensions are specified using the upper and lower deviations. In case of a fit, the basic size of both connected elements must be the same.

The tolerance of a basic size is defined as the difference between the upper and lower limit dimensions of the part. In order to meet the requirements of various production branches for accuracy of the product, the system ISO implements 20 grades of accuracy. Each of the tolerances of this system is marked "IT" with attached grade of accuracy (IT01, IT0, IT1, …, IT18).

The tolerance zone is defined as an annular zone limited by the upper and lower limit dimensions of the part. The tolerance zone is therefore determined by the amount of the tolerance and its position related to the basic size. The position of the tolerance zone, related to the basic size (zero line), is determined in the ISO system by a so-called basic deviation. The system ISO defines 28 classes of basic deviations for holes. These classes are marked by capital letters (A, B, C, …, ZC). The tolerance zone for the specified dimensions is prescribed in the drawing by a tolerance mark, which consists of a letter marking of the basic deviation and a numerical marking of the tolerance grade (e.g. H7, H8, D5, etc.). Though the general sets of basic deviations (A, B, C, …, ZC) and tolerance (IT01, IT0, IT1, …, IT18) can be used for prescriptions of hole tolerance zones by their mutual combinations, in practice only a limited range of tolerance zones is used. The system ISO also defines 28 classes of basic deviations for shafts. These classes are marked by lower case letters (a, b, c,⋯, zc). The tolerance zone for the specified dimensions is prescribed in the drawing by a tolerance mark, which consists of a letter marking of the basic deviation and a numerical marking of the tolerance grade (e. g. h7, h6, g5, etc.). Though the general sets of basic deviations(a, b,⋯, zc) and tolerance grades (IT1, IT2,⋯, IT8) can be used for prescriptions of shaft tolerance zones by their mutual combinations, in practice only a limited range of tolerance zones is used.

Although there can be generally coupled parts without any tolerance zones, only two methods of coupling of holes and shafts, namely basic hole system and basic shaft system, are recommended due to constructional, technological and economic reasons.

Any manufacturer should have a thorough of the tolerances and fits to increase the quality and reliability of a manufactured part with the least expense. The increased compatibility of a certain product with mating parts made by different manufacturers makes it more viable in market competition. Moreover, a knowledge of tolerances and fits is essential for manufacturing with least material wastage, and hence good tolerance principles followed by manufacturers are essential in reaping profits.

New Words and Expressions

1. ISO　国际标准化组织
2. smooth part　光滑零件
3. cylindrical part　圆柱体零件
4. round section　圆截面
5. upper deviation　上偏差
6. lower deviation　下偏差

Lesson 9 Numerical Control

Text

Today, conventional machine tools have been largely replaced by Computer Numerical Control(CNC) machine tools. The machines still perform essentially the same functions, but movements of the machine tool are controlled electronically rather than by hand. CNC machine tools can produce the same parts over and over again with very little variation. They can run day and night, week after week, without getting tired. These are obvious advantages over conventional machine tools, which need a great deal of human interaction in order to do anything.

This is not to say that conventional machine tools are obsolete. They are still used extensively for tool and fixture work, maintenance and repair, and small volume production. However, much of the high-and medium-volume production work is now performed on CNC machine tools. Furthermore, CNC machining has become common in the low-volume job shops where lot sizes of a dozen to several hundred parts are common.

CNC machine tools are highly productive. They are also expensive to purchase, set up, and maintain. However, the productivity advantage can easily offset this cost if their use is properly managed. The decision to produce parts conventionally or by CNC is driven mainly by setup cost and volume. The setup cost to perform a machining operation on a conventional machine tool is rather low; on a CNC machine tool, the setup cost can be rather high. In summation, as the production volume increases, the combination of productivity gains and the ability to spread the setup cost over many parts makes CNC machining the obvious choice.

CNC machine tools are complex assemblies, and a more detailed study is a topic for a separate book. However, in general, any CNC machine tool consists of the following units: Computers, Control systems, Drive motors, Tool changers.

According to the construction of CNC machine tools(see Figure 9.1), CNC machines work in the following manner:

(1) The CNC machine control computer reads a prepared program and translates it into machine language, which is a programming language of binary notation used on computers, not on CNC machines.

(2) When the operator starts the execution cycle, the computer translates binary codes into electronic pulses which are automatically sent to the machine's power units. The control units compare the number of pulses sent and received.

(3) When the motors receive each pulse, they automatically transform the pulses into rotations that drive the spindle and lead screw, causing the spindle rotation and slide or table

movement. The part on the milling machine table or the tool in the lathe turret is driven to the position specified by the program.

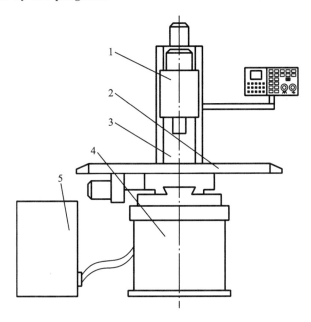

Figure 9.1 CNC machine tools

1—headstock; 2—table; 3—colum; 4—bed; 5—electrical control system of the machine

There also have some related concepts. Such as numerical control (NC) and machining center.

Numerical control (NC) is a form of programmable automation in which the processing equipment is controlled by meas of numbers, letters, and other symbols. The numbers, letters, and symbols are coded in an appropriate format to define a program of instructions for a particular work-part or job. When the job changes, the program of instructions is changed.

The NC system consists of the following components: data input, the tape reader with the control unit, feedback devices, and the metal-cutting machine tool or other type of NC equipment.

Data input, also called "man-to-control-link", may be provided to the machine tool manually, or entirely by automatic mean. Manual methods when used as the sole source of input data are restricted to a relatively small number of inputs. Manual input requires that the operator set the controls for each operation. It is a slow and tedious process and is seldom justified except in elementary machining applications or in special cause.

While the data on the tape is fed automatically, the actual programming steps are done manually. Before the coded tape may be prepared, the programmer, often working with a planner or a process engineer, must select the appropriate NC machine tool, determine the kind of material to be machined, calculate the speeds and feeds, and decide upon the type of tooling needed. The dimensions on the part print are closely examined to determine a suitable zero reference point from which to start the program, manuscript is then written which gives coded numerical instructions describing the sequence of operations that the machine tool is required to follow to cut the part to the drawing specification.

The control unit receives and stores all coded data until a complete block of information has been accumulated. It then interprets the coded instruction and directs the machine tool through the required motion.

Silicon photo diodes, located in the tape reader head on the control unit, detect light as it passes through the holes in the moving tape. The light beams are converted to electrical energy, which is amplified to further strengthen the signal. The signals are then sent to registers in the control unit, where actuation signals are relayed to the machine tool drives.

A machining center can be defined as a machine tool capable of:
(1) Multiple operation and processes in a single set-up utilizing multiple axis;
(2) Typically has an automatic mechanism to change tools;
(3) Machine motion is programmable;
(4) Servo motors drive feed mechanisms for tool axis;
(5) Positioning feedback is provided by resolver to the control system.

There are two main types of machining centers: the horizontal spindle and the vertical spindle machine.

Horizontal spindle type　The traveling-column type is equipped with one or usually two tables on which the workpiece can be mounted. With this type of machining center, the workpiece can be machined while the operator is loading a new workpiece on the other table.

The fixed-column type is equipped with a pallet shuttle. The pallet is a removable table on which the workpiece is mounted. After the workpiece has been machined, the workpiece and pallet are moved to a shuttle which then rotates, bringing a new pallet and a new workpiece into position for machining.

Vertical spindle type　The vertical spindle machining center is a saddle-type construction with sliding bedways, which utilizes a sliding vertical head instead of a spindle movement.

New Words and Expressions

1. computer numerical control (CNC) 计算机数字控制
2. productivity 生产率, 产量
3. variation 变化, 变动
4. obsolete 过时的
5. small volume production 小批量生产
6. conventional 普通的, 常规的
7. spindle 轴
8. screw 丝杠, 螺杆
9. numerical control(NC) 数字控制
10. machining center 加工中心
11. servo motor 伺服电动机
12. horizontal 卧式的
13. vertical 立式的
14. pallet 托盘
15. shuttle 梭子, 往复移动送件装置
16. saddle 滑板
17. bedway 导轨

Notes

(1) In summation, as the production volume increases, the combination of productivity gains and the ability to spread the setup cost over many parts makes CNC machining the obvious choice.

总的来说,由于生产量增加,生产率提高,配置成本分散到众多部件上,数控加工成为显而易见的选择。

(2) The pallet is a removable table on which the workpiece is mounted. After the workpiece has been machined, the workpiece and pallet are moved to a shuttle which then rotates, bringing a new pallet and a new workpiece into position for machining.

托盘是一个可移动的平台,用来安装工件。工件加工完后,工件和托盘被移动到往复移动送件装置中轮转,将新的托盘和新工件送至加工位置。

❖ Translation Skills

否定结构的翻译

英语的否定形式是常见而又比较复杂的,使用非常灵活、微妙,被认为是英语的特点之一。科技英语里否定形式的应用也很广泛。在表达否定概念时,英语在用词、语法和逻辑等方面与汉语有很大的不同。有的句子形式上是肯定的而事实上是否定的,而有的句子形式上是否定的但事实上却是肯定的。英语否定词的否定范围和重点有时难以判断,否定词在句子中表达强调的方法与汉语不相同,某些否定词和词组的习惯用法较难掌握。因此,在翻译英语否定形式时,必须细心揣摩,真正彻底理解其意义否定的重点,然后根据汉语的习惯进行翻译。

[例1]　The earth does not move round in the empty space.

译文:(误)地球在空无一物的空间中不运转。

　　　(正)地球不是在空无一物的空间中运转。

[例2]　All metals are not good conductors.

译文:(误)所有金属都不是良导体。

　　　(正)并非所有金属都是良导体。

从上面句子中可以看出,英语的否定形式与否定概念并非永远一致。例1那样的句子,形式上是一般否定(谓语否定),但实际上确实特指否定(其他成分否定);如例2,句子看上去似乎是全部否定,但却是部分否定。下面我们就针对翻译英语否定结构时应注意的几个问题,举例加以说明。

一、否定成分的转译

一般来说,否定形式仍翻译成否定形式。但由于英语和汉语两种语言其表达手段和习惯不同,有些否定句应译成肯定形式。这种正反、反正表达法是翻译的一门重要技巧。有些英语否定句,虽然是用一般否定(否定谓语)的形式,但在意义上却是特指否定,即其他成分的否定;反之,有些句子形式上是特指否定,而意义上却是一般否定。翻译时,要根据汉语的习惯进行否定成分的翻译。

［例1］ Sound does not travel so fast as light.

译文：声音不像光传播得那样快。（原文否定谓语，译文否定状语）

［例2］ Neutrons carry no charge.

译文：中子不带电荷。（原文否定宾语，译文否定谓语）

［例3］ The sun's rays do not warm the water so much as they do the land.

译文：太阳光线使水增温，不如它使陆地增温那样高。（原文否定谓语，译文否定状语）

［例4］ Matter must move, or no work is done.

译文：物质必须运动，否则就不做功。（原文否定主语，译文否定谓语）

［例5］ The mountains is not valued because it is high.

译文：山的价值并不是因为它高。（原文否定谓语，译文否定谓语）

［例6］ Nobody can be set in motion without having a force act upon it.

译文：如果不让力作用在物体上，就不能使物体运动。（原文否定主语，译文否定谓语）

［例7］ Green plants cannot grow strong and healthy without sunlight.

译文：没有阳光，绿色植物就长不结实，长不好。

［例8］ But for the heat of the sun, nothing could live.

译文：要是没有太阳的热，什么东西都不能生存。（原文否定主语，译文否定谓语）

［例9］ We do not consider melting or boiling to be chemical changes.

译文：我们认为熔化和沸腾不是化学反应。（原文否定谓语，译文否定兼语式的第二谓语）

二、部分否定结构的翻译

英语的否定有全部否定与部分否定之分。全部否定指否定整个句子的全部意思，可用 none, neither, no, not, nothing, nobody 等否定词。部分否定则主要由 all, every, both, always 等含全体意义的词与否定词 not 构成，其表达的意义是部分否定，相当于汉语"不是所有都"、"不是两者都"、"不总是"之意。当否定词 not 放在这些词(not all, not every, not both)之前时，其部分否定的意义就很明显，一般不会翻译错。然而，否定词 not 有时却与谓语在一起，构成谓语否定，形式上很像全部否定，但实际上却是部分否定，翻译中应当特别注意。

［例1］ All minerals do not come from mines. (＝Not all minerals come from mines.)

译文：并非所有矿物都来自矿山。

［例2］ Every color is not reflected back. (＝Not every color is reflected back.)

译文：并非每种色光都会反射回来。（不是"每种色光都不反射回来"。）

［例3］ Both of the substances do not dissolve in water.

译文：不是两种物质都溶于水。

［例4］ But friction is not always useless, in certain cases it becomes a helpful necessity.

译文：摩擦并非总是无用的，在某些场合下，它是有益的、必需的。

［例5］ All the chemical energy of fuel is not converted into heat.

译文：并非所有燃料的化学能都转变成热量。

三、否定语气的改变

英语的否定句不能一概译成汉语的否定句，因为英语中有些否定句表达的是肯定的意思，还有些否定句在特定场合下可以表达肯定的意思。试看下面带否定词 nothing 的句子的译法。

［例1］ Energy is nothing but the capacity to do work.

译文：能就是做功能力。

[例2] An explosion is nothing more than a tremendously rapid burning.

译文：爆炸仅仅是非常急速的燃烧。

[例3] Ball bearings are precisely made bearings which make use of the principle that "nothing rolls like a ball".

译文：滚珠轴承是精密轴承，它采用了"球最善于滚动"的原理。

四、否定意义的表达

英语中还有许多肯定句，其所表达的是否定的意思。在这类句子中虽然没有出现否定词，但句子中有些词组却含有否定的意义。翻译时一般都将其否定意义译出，译成汉语的否定句。对于这类句子，只要掌握了词汇的意义，理解和翻译就不会有什么困难。在科技英语中，常见的含有否定意义的词组有 too...to(太……不)，free from(没有)，too...for(太……不)，fall short of(没有达到)，fail to(不能)，instead of(而不是)，far from(完全不)，in the absence of(没有……时)。

[例1] Of all metals silver is the best conductor, but it is too expensive to be used in industry.

译文：在所有金属中，银是最好的导体，但它太昂贵，不能在工业上使用。

[例2] The distance from the sun to the earth is too great to imagine.

译文：太阳到地球的距离远得不可想象。

[例3] The angularity of the parts is too great for proper assembly.

译文：零件的斜度太大，不适于装配。

[例4] Hardened steel is too hard and too brittle for many tools.

译文：对许多刀具来说，淬火钢太硬、太脆，都不能用它来制造。

[例5] If the follower loses contact with the cam, it will fail to work.

译文：随动元件如果与凸轮脱开，就不能工作。

五、双重否定结构的翻译

英语和汉语一样，也有双重否定结构。英语的双重否定结构是由两个否定词(no, not, never 等)连用或一个否定词与某些表示否定意义的词连用而构成的。双重否定表示否定之否定，即强调肯定，因此双重否定结构有两种译法，既可以译成双重否定结构，也可以译成肯定结构。

双重否定从语气的强弱上分，有弱化的双重否定和强化的双重否定两种。弱化的双重否定中包含的两个否定词，一个常由否定前、后缀构成，由于两个否定词中一个为另一个所否定，使否定的语气弱化，从而将否定的效果抵消一部分。

(一) 译成双重否定结构

[例1] There is no steel not containing carbon.

译文：没有不含碳的钢。

[例2] Sodium is never found uncombined in nature.

译文：在自然界中从未发现不处于化合状态的钠。

[例3] No flow of water occurs unless there is difference in pressure.

译文：没有压力差，水就不会流动。

[例4] It is impossible for heat to be converted into a certain energy without something lost.

译文:热转换成某种能而没有什么耗损是不可能。

(二) 译成肯定结构

[例1]　There is no law that has not exceptions.

译文:凡是规律都有例外。

[例2]　A radar screen is not unlike a television screen.

译文:雷达荧光屏跟电视荧光屏一样。

[例3]　There is nothing unexpected about it.

译文:一切都在意料之中。

[例4]　One body never exerts a force upon another without the second reacting against the first.

译文:一个物体对另一物体施加作用力,就必然会受到另一物体的反作用力。

❖ Reading Material

Numerical Control and Automatic Machines

The major disadvantage of machine tool lies in the economics of the process. It is expensive to a machine tool for automatic production. Therefore, unless the part is to be made in very large numbers, the cost becomes prohibitive. Great need exists for a method that permits rapid automatic production, economical lot amounts in job. The answer has been found in the numerical control of machine tools. In numerical control, the blueprint for a part is converted into a punched paper type instruction, which is adapted via a computer to direct the operation of a specific machine tool. Thus general purpose machine tools described in previous chapters are instructed to machine a part according to information stored on a roll of tape. The tape can be rerun for copies of the same part or can be stored for future use. Furthermore, other tapes can be used to command the same machine tool to make other parts. There is a wide area of performance duplication between numerical control and automatics. Numerical control, however, offers more flexibility, lower tooling cost, quicker changes, and less machine down time.

In machining contours, numerical control can mathematically translate the defined curve into a finished product, saving time and eliminating templates. This can in turn improve accuracy. Another advantage appears to be great saving of machine time, the equivalent of increasing productive capacity with no increase in facilities.

Automation has no new development. Semiautomatic machines have been used in the textile industry and in engineering for many years. These machines merely require to be set up, loaded and started. Then, for a limited time, they will run on their own, with only an operator to watch them. From these have been developed machines known as transfer machines, found mostly in the motor manufacturing industry. In production units using these, each stage in the manufacture of an article is carried out by one fully automatic machine in a line of machines. The loading of the article to be machined is automatic, as it is transfer from one machine to the next.

Most of the transfer machines currently in operation employ electrical, pneumatic or hydraulic techniques.

Although automatic control by pneumatic and hydraulic means has been developed to a high degree of efficiency, the more recently developed electronic techniques offer many advantages over them. Electronic methods allow for greater speed, accuracy and flexibility in the operation of control systems. They also make possible the processing of information. Information, both from within the control system itself and from outside sources, can be electronically processed. Thus the control of extremely complex processes can be carried out automatically.

New Words and Expressions

1. instruction   指令,命令
2. via   通过,经由
3. duplication   复制,副本
4. pneumatic   气动的,充气的
5. hydraulic   液压的,水力的

Lesson 10 Material Forming

Text

The term material forming refers to a group of manufacturing methods by which the given shape of a work piece is converted to another shape without change in the mass or composition of the material of the work piece. The stresses induced during the process are greater than the yield strength, but less than the fracture strength, of the material. The type of loading may be tensile, compressive, bending, or shearing, or a combination of these. This is a very economical process as the desired shape, size, and finish can be obtained without any significant loss of material. Moreover, a part of the input energy is fruitfully utilized in improving the strength of the product through strain hardening.

1. Cold forming and hot forming

The forming processes can be grouped under two broad categories, namely, cold forming, and hot forming. If the working temperature is higher than the recrystallization temperature of the material, then the process is called hot forming. Otherwise the process is termed as cold forming.

Many techniques are used to simultaneously shape and strengthen a material by cold working. For example, rolling is used to produce metal plate and sheet. Forging deforms the material into a die cavity, producing relatively complex shapes such as automotive crankshafts or connecting rods. Deep drawing is used to form the body of aluminum beverage cans. In extrusion, a material is pushed through a die to form products of uniform cross-sections, including rods, tubes, or aluminum trims for doors or windows. Stretch forming and bending as well as other processes are used to shape material. Thus cold working is an effective way of shaping metallic materials while simultaneously increasing their strength. We can obtain excellent dimensional tolerances and surface finishes by the cold working processes.

Hot working is well-suited for forming large parts, since the metal has a low yield strength and high ductility at elevated temperatures. In addition, HCP metals such as magnesium have more slip systems at hot-working temperature; the high ductility permits larger deformation than are possible by cold working. For example, a very thick plate can be reduced to a thin sheet in a continuous series of operations. An advantage of hot working is that the imperfection of materials can be eliminated during the process. Some imperfection in the original material maybe eliminated or their effects minimized; gas pores can be closed and welded shut; composition differences in the metal can be reduced; the structure of metals can be refined and controlled by recrystallization. Therefore, mechanical and physical properties of metals can be improved obviously.

The typical forming processes are rolling, forging, deep drawing and bending. For a better understanding of the mechanics of various forming operations, we shall briefly discuss each of the process.

2. Rolling

Rolling can be characterized as: mass conserving, solid state of material, mechanical primary basic process-plastic deformation. Rolling is extensively used in the manufacturing of plates, sheets, structural beams, and so on. Figure 10.1 shows the rolling of plates or sheets. An ingot is produced in casting and in several stages it is reduced in thickness, usually while hot. Since the width of the work material is kept constant, its length is increased according to the reductions. After the last hot-rolling stage, a final stage is carried out cold to improve surface quality and tolerances and to increase strength. In rolling, the profiles of the rolls are designed to produce the desired geometry.

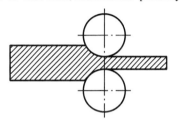

Figure 10.1 Rolling operation

3. Forging

Forging is one of the most important manufacturing operations. It is a plastic deformation process similar to extrusion but, unlike extrusion, it can be used to manufacture complex 3D parts. It is used in producing components of all shapes and sizes, from quite small items to large units weighing several tons. Forging can be classified into two main categories: open-die forging, closed-die forging.

Open-die forging The process is schematically illustrated in Figure 10.2(a). At least one of the work piece surfaces deforms freely, and hence the open-die forging process produces parts of lesser accuracy and dimensional tolerance than closed-die forging. However, the tooling is simple, relatively inexpensive and can be designed and manufactured with ease.

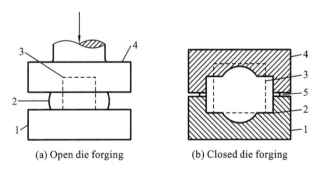

(a) Open die forging   (b) Closed die forging

Figure 10.2 Forging operation

1—die1; 2—final shape; 3—original shape; 4—die 2; 5—flash

Closed-die forging In closed-die forging the work piece is completely trapped in the die and no flash is generated. Material utilization is very high, but the volume of the work piece before and after forging is identical and hence control of incoming material volume becomes critical. Excess material can create large pressures, which are liable to cause die failure.

4. Deep drawing

In deep drawing, a cup-shaped product is obtained from a flat sheet metal with the help of a punch and a die. Figure 10.3 shows the operation schematically. The sheet metal is held over the die by means of a blank holder to avoid defects in the product.

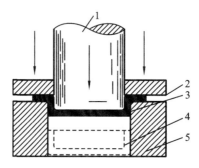

Figure 10.3　Deep drawing
1—punch;2—blank holder;3—work piece;4—final product;5—die

A great variety of parts are formed by this process, the successful operation of which requires a careful control of factors such as blank-holder pressure, lubrication, clearance, material properties, and die geometry. Depending on many factors, the maximum ratio of blank diameter to punch diameter ranges from about 1.6 to 2.3.

This process has been extensively studied, and the results show that two important material properties for deep draw ability are the strain-hardening exponent and the strain ratio (anisotropy ratio) of the metal. The former property becomes dominant when the material undergoes stretching, while the latter is more pertinent for pure radial drawing. The strain ratio is defined as the ratio of the true strain in the width direction to the true strain in the thickness direction of a strip of the sheet metal. The greater this ratio, the greater is the ability of the metal to undergo change in its width direction while resisting thinning.

Anisotropy in the sheet plane results in crease, the appearance of wavy edges on drawn gasket. Clearance between the punch and the die is another in this process; this is normally set at a value of not more than 1.4 times the thickness of the sheet. Too large a clearance produces a gasket whose thickness increases toward the vertical direction, whereas correct clearance produces a gasket of uniform thickness by ironing. Also, if the blank-holder pressure is too low, the flange wrinkles; if it is too high, the bottom of the gasket will be punched out because of the increased frictional resistance of the flange. For relatively thick sheets, it is possible to draw parts without a blank holder by special die designs.

5. Bending

As the name implies, this is a process of bending a metal sheet plastically to obtain the desired shape. This is achieved by a set of suitably designed punch and die. A typical process is shown schematically in Figure 10.4.

Bending and forming may be performed on the same equipment as that used for shearing—namely, crank, eccentric, and cam-operated presses. Where bending is involved the metal is stressed in both tension and compression at values below the ultimate strength of the material without appreciable changes in its thickness. As in a press brake, simple bending implies a straight bend across the sheet of metal. Other bending operations, such as curling, seaming, and folding, are similar, although the process is slightly more improved.

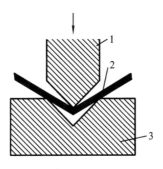

Figure 10.4　Bending
1—punch; 2—workpiece; 3—die

New Words and Expressions

1. yield strength　屈服强度
2. fracture strength　断裂强度
3. recrystallization temperature　再结晶温度
4. beverage can　饮料罐
5. crankshaft　曲柄轴
6. ductility　延性
7. slip system　滑移系
8. imperfection　不完美，瑕疵

Notes

(1) The stresses induced during the process are greater than the yield strength, but less than the fracture strength, of the material. The type of loading may be tensile, compressive, bending, or shearing, or a combination of these.

在此工艺过程中，材料产生的应力大于材料的屈服强度，但小于材料的断裂强度。应力的类型可以是拉应力、压应力、弯曲应力或切应力，或者是这些类型应力的组合。

(2) An advantage of hot working is that the imperfection of materials can be eliminated during the process. Some imperfection in the original material maybe eliminated or their effects minimized; gas pores can be closed and welded shut; composition differences in the metal can be reduced; the structure of metals can be refined and controlled by recrystallization.

热加工的优点是通过加工能消除材料中的缺陷：可消除原材料中的一些缺陷或将其不良影响减至最少；气孔能被压合或熔合，材料中的成分偏析会减少；通过再结晶可以很好地细化及控制金属的显微组织。

❖ Translation Skills

长句的译法

科技英语的特点主要是术语繁杂、论证严密，翻译时要特别注意逻辑是否严谨、术语是否准确、表达是否简练和明确。不仅并列句和复合句，简单句有时也很长，它的一个重要特点就是修饰语较长，这些修饰语一般都是短语和从句，它们或是位于名词后面的短语或从句，或是位于动词后面的短语或从句。因此，为了兼顾译文的准确性和可读性，我们就要掌握一定的方法和技巧，这些技巧包括顺译法、逆译法、分译法和综合法等。

一、顺译法

[例 1] （1）Fossil fuel emits another 5.2 billion metric tons of CO_2 into the air each year, (2) while the burning of tropical forests emits roughly 1.8 billion metric tons of CO_2——(3) both contributing to a build up of carbon dioxide (4) that will soon trigger the greenhouse effect.

分析：在这个句子中，主句是(1)，(2)是主句的并列句，(3)是同位语，(4)是定语从句。顺句翻译得到的译文不难理解，比较符合汉语的表达习惯，适合采用顺译法。

译文：矿物燃料每年向大气释放52亿吨二氧化碳，同时热带森林的燃烧大约释放出18亿吨二氧化碳——这两方面都导致了二氧化碳的聚集，因而会很快引发温室效应。

二、逆译法

有时候英语长句的叙述层次与汉语相反，这时就需从英语原文的后面译起，逆着英语原文的顺序翻译。

[例 1] (1) There is an equilibrium between the liquid and its vapor, (2) as many molecules being lost from the surface of the liquid and then existing as vapor, (3) as reenter the liquid (4) in a given time.

分析：这个句子分为四层意思：(1)"液体和蒸汽之间处于平衡"，是主句，是全句的中心所在；(2)"许多分子从液体表面溢出，成为蒸汽"；(3)"又有同样多的分子重新进入液体"；(4)在一定时间内。根据汉语表达习惯，"因"在前面，"果"在后，可逆着原文翻译。

译文：在一定时间内，许多分子从液体表面逸出成为蒸汽，又有同样多的分子重新进入液体。因此，液体和蒸汽之间处于平衡。

[例 2] (1) Rocket research has confirmed a strange fact (2) which had already been suspected, (3) there is a "high-temperature belt" in the atmosphere, with its center roughly 30 miles above the ground.

分析：(1)是主句；(2)是定语从句；(3)是同位语从句，做 fact 的同位语。按照汉语表达习惯，用逆译法翻译，更显得自然、流畅。

译文：大气中有一个"高温带"，其中心在距地面约30英里（1英里＝1609.344千米）高的地方。对此人们早就有怀疑，利用火箭加以研究，使这件奇怪的事情得到了证实。

三、分译法

有时英语长句中主句与从句或主句与修饰语间的关系不十分密切，翻译时可按照汉语多用短句的习惯，把长句中的从句或短语化为句子，分开来叙述，将原句化整为零。为使译文通顺连贯,也可以适当加几个连接词。

[例1] The Josephson junction is a thin insulator separating two superconducting materials, i. e. , metals that lose all electrical resistance at temperature near zero.

译文:约瑟生接点就是分开两种超导体材料的薄绝缘体。超导材料是在接近绝对零度时会失去电阻的金属。

[例2] It is very important that a machine element be made of a material that has properties suitable for the condition of service as it is for the loads and stresses to accurately determined.

译文1:重要的是机器零件要用性能符合工作条件的材料制造。同样重要的是,准确计算出载荷和压力。(分译法)

译文2:载荷和压力必须准确地计算出来,机器零件要用性能符合工作条件的材料来制造。这两件事都是非常重要的。(逆译法)

本例也说明,技巧不是绝对的,只要译文符合汉语表达习惯,分译法和逆译法都可以采用。

四、综合法

如果利用上述技巧无法翻译出可读性好的译文,译者可以着眼于整体,以逻辑分析为基础,分清主次,对全句进行综合处理。

[例1] Computer languages may range from detailed low level close to that immediately understood by the particular computer, to the sophisticated high level which can be rendered automatically acceptable to a wide range of computers.

译文:计算机语言有低级的也有高级的。前者比较烦琐,类似某种特定的计算机能直接理解的语言;后者比较复杂,适用范围广,很多计算机能自动接受。

[例2] Applied science, on the other hand, is directly concerned with the application of the working laws of pure science to the practical affairs of life, and to increasing man's control over his environment, thus leading to the development of new techniques, processes and machines.

译文:另一方面,应用科学则直接涉及应用问题。它研究的是如何将纯科学的工作定律应用于实际生活、应用于加强人类对其周围环境的控制,从而促使新技术、新工艺和新机器产生。

上述译文综合归纳出了句子的中心思想,简单明了,读者很容易获取到有用信息。

❖ Reading Material

Types of Bending

The basic types of bending applicable to sheet metal forming are straight bending, flange bending, contour bending(see Figure 10.5), stretch bending. Examples of these four types of bending are showed:

Straight bending The terminology for a straight bend is shown in Figure 10.6. During the forming of a straight bend the inner grains are elongated in the bend zone.

Tensile strain builds up in the outer grains and increases with the decreasing bend radius. Therefore, the minimum bend radius is an important quantity in straight bending

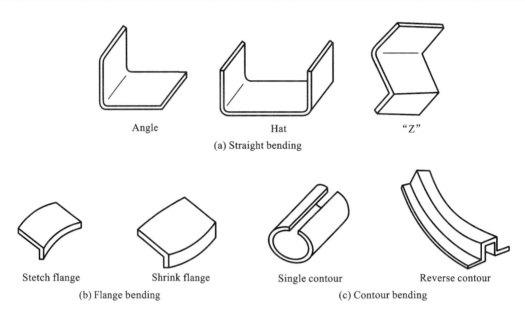

Figure 10.5 Types of bend forming

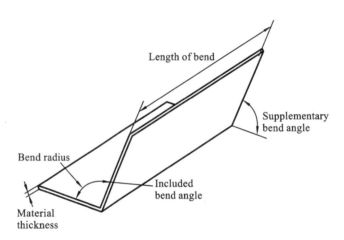

Figure 10.6 Terminology for a straight bend

beyond which splitting occurs.

Flange bending Flange bend forming consists of forming shrink and stretch flanges as illustrated by views in Figure 10.5(b). This type of bending is normally produced on a hydrostatic or rubber-pad press at room temperature for materials such as aluminum and light-gage steel.

Parts requiring very little handwork are produced if the flange height and free-form radius requirements are not severe. However, forming metals with low modulus of elasticity to yield strength ratios, such as magnesium and titanium, may result in undesirable buckling and springback as shown in Figure 10.7(a) and (b). Also, splitting may result during stretch flange forming as a function of material elongation.

Elevated temperatures utilized during the bending operation enhance part formability and definition by increasing the material ducting and lowering the yield strength, providing less

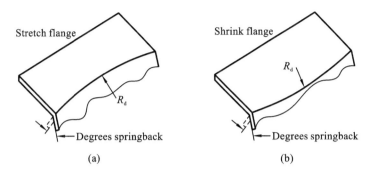

Figure 10.7　Flange forming failure modes

spring back and bucking.

Contour bending　Contour bending is illustrated by the single contoured part, and the reverse contoured part. Single contour bending is performed on a three-roll bender or by using special feeding devices with a conventional press brake. Higher production rates are attained using a three-roll bending machine. Contour radii are generally quite large; forming limits are not a factor. However, spring-back is a factor because of the residual-stress buildup in the part; therefore, overforming is necessary to produce a part within tolerance.

Stretch bending　Stretch bending is probably the most sophisticated bending method and requires expensive tooling and machines. Further more, stretch bending requires lengths of material beyond the desired shape to permit gripping and pulling. The material is stretched longitudinally, past its elastic limit by pulling both ends and then wrapping around the bending form. This used primarily for bending form irregular shapes; it is generally not used for high production.

New Words and Expressions

1. flange　边缘,轮缘,凸缘,法兰
2. rubber-pad　橡胶垫
3. buckling　变形,结构失稳,翘曲
4. spring-back　回弹,弹性后效
5. residual-stress　残余应力,残留应力

Lesson 11 Flexible Manufacturing

Text

Flexibility refers to producing a number of distinct product types as in low volume job shop production, while productivity implies high-speed production that is similar to automated high volume mass production. In general, increasing productivity and maintaining flexibility have been one of major goals in most industry sectors. Emergence of the flexible manufacturing system (FMS) lies in this direction. In other words, the FMS attempts to achieve both flexibility and productivity while meeting the fluctuating demands occurred in today's competitive market. Over the last decades, the FMS has gone through rapid evolution and development, and many effective systems have been installed in a wide range of manufacturing environment. FMS is defined as a manufacturing system with computer-controlled workstations and a material handling system designed to manufacture more than one kind of part type in low to medium volumes to meet customer requirements. Basically, FMS is a combination of machine tools, material handling equipments, and computer software as indicated in Figure 11.1. Various types of FMSs can be found in the industry. The most representatives are flexible machining cell, flexible manufacturing system, and flexible transfer line.

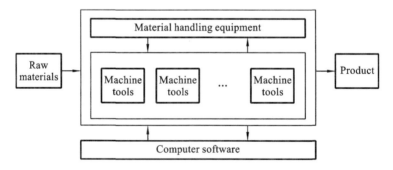

Figure 11.1 Basic elements of a FMS

The key components of a flexible manufacturing system include:

(1) Processing machines arranged in line with the part processing sequence. These machines are usually CNC (computer numerical control) machines that perform processing operations on certain parts and components. Other types of automated workstations such as inspection machines may also be included in a flexible manufacturing system.

(2) An automated material-handling system, such as an overhead automatic loading and unloading system combined with conveyors for complete transport of parts and components

throughout the flexible manufacturing system. The material handling system is also electronically inter connected with all the machines for electronic information exchange for part loading and unloading.

(3) A central computer system that is responsible for the inter connection and information exchange between each machine and the material handling system, coordinating the movement of the gripers in the automatic material handling system for timely loading and unloading of work parts. Part machining or processing programs are pre loaded into each machine in the flexible manufacturing system for production of different types of parts.

A traditional flexible manufacturing system (FMS) is an arrangement of machines interconnected by a part transport system. In the early days of application in the manufacturing industry, flexible manufacturing systems were often applied in the design and configuration of job shops where different types of parts and components can be processed in a small batch size. The key benefit from an early flexible manufacturing system is the capability to process more than one type of part with the same set of machines and equipment. Those systems were ideal for part suppliers serving many customers with low to medium quantity but large part variety.

With the development of computer science and its application in the industry, advanced processing control technologies, such as programmable logical control (PLC) and computer numerical control (CNC) were developed and applied in the manufacturing industry since the 1970s up to the present time. The need for high volume production with higher productivity and efficiency, as well as the need for the capability to produce different parts without major retooling cost, have promoted the design of much more efficient flexible manufacturing systems with automatic control and material handling capability. These modern flexible manufacturing systems were used in major manufacturing plants in the automotive industry and other industries around the world. They were well designed for medium and high volume production with large batch size, with the flexibility to change quickly from one type of product to another.

Because of the cost benefits and operational advantages generated with flexible manufacturing systems, companies in the automotive industry, including original vehicle manufacturers and part and component suppliers, have embraced the flexible manufacturing concept and have successfully implemented flexible manufacturing systems in their production facilities in the United States and in their overseas operations. Due to the competitiveness of the automotive industry and the large population of motor vehicle ownership in the United States and in other countries around the world, production systems in the automotive industry have to be designed and managed in such a way that allows for high volume production with high efficiency, yet makes it possible for a quick change of product type in response to changes in market demand without incurring too much retooling cost. Well designed and managed flexible manufacturing systems, could be better solutions for production in the automotive industry than dedicated production lines. Companies which successfully implemented flexible manufacturing strategy are able to achieve the long term benefits from high volume production for the economy of scale and wide variety of production for the economy of scope. Companies doing well in both aspects will have a better perspective for sustained business growth and success in the long run.

The advantages of a flexible manufacturing system may include:
- Faster, lower-cost changes from one part to another which will improve capital utilization.
- Lower direct labor cost, due to the reduction in number of workers.
- Reduced inventory, due to the automated control.
- Lower cost per unit of output, due to the greater productivity using the same number of workers.
- Saving from the indirect labor, reduced errors, rework, repairs, and rejects.

Manufacturing operations are inevitably faced with wide range of uncertainties and variations in production process. Companies need to handle them in advance or react after their occurrence. Uncertainty and unexpected events may change the system status and affect the performance. The production schedule employed as a crucial tool in manufacturing systems in order to increase productivity and decrease the operating cost, is subject to be upset widely by disruptions during the execution schedules. Machines may be stopped during their operations for major failures such as breakdown, toolset wearing, or tool reassignments. If these disruptions cause significant deterioration in performance, the system needs to react and update the existing schedule in order to lessen the impact. The main disadvantages are:
- Limited ability to adapt to changes in product or product mix.
- Substantial pre-planning activity.
- Relatively high cost.
- Technological problems of exact component positioning and precise timing.
- Sophisticated manufacturing system.

New Words and Expressions

1. flexible manufacturing 柔性制造
2. inventory 存货,存货清单
3. configuration 结构,配置
4. material handling 物料运输
5. technological 技术的,工艺的
6. automatic 自动化的
7. sophisticated 复杂的,精致的
8. programmable logical control 可编程控制器(PLC)
9. assembly plant 装配厂
10. in response to 响应,回答,对……有反应
11. pick-up truck 客货两用车
12. capital 资金;资本家;重要的
13. derivative 派生的,引出的
14. up to 一直到

Notes

(1) Over the last decades, the FMS has gone through rapid evolution and development, and many effective systems have been installed in a wide range of manufacturing environment.

在过去的几十年中,柔性制造系统经历了快速的演变和发展,并且在一些制造企业中已经实现了许多行之有效的系统。

(2) FMS is defined as a manufacturing system with computer-controlled workstations and a material handling system designed to manufacture more than one kind of part type in low to medium volumes to meet customer requirements.

FMS 定义为一种由计算机控制的工作站和物料处理系统组成,用来进行多品种、中小批量零件生产以满足客户需要的制造系统。

(3) In the early days of application in the manufacturing industry, flexible manufacturing systems were often applied in the design and configuration of job shops where different types of parts and components can be processed in a small batch size.

在制造业发展的初期,柔性制造系统通常应用在加工车间的设计和布置方面,在这些加工车间可以生产小批量、不同类型的零部件。

(4) Because of the cost benefits and operational advantages generated with flexible manufacturing systems, companies in the automotive industry, including original vehicle manufacturers and part and component suppliers, have embraced the flexible manufacturing concept and have successfully implemented flexible manufacturing systems in their production facilities in the United States and in their overseas operations.

由于柔性制造系统具有成本效益和运营优势,美国汽车行业中包括原始的汽车制造商和零部件供应商在内的一些企业已经接受柔性制造的概念,并且已经在其国内外的企业中成功地使用了柔性制造系统。

❖ Translation Skills

科技文章的特点

科技英语是一种信息语言体,是科技人员借以传递科技信息的媒介,其主要内容要求客观、真实,具有较强的科学性,因此在语法结构上应区别于文学英语。文学英语主要特点是再现性和表现性的高度统一,语言修饰和节奏变幻多样,而科技英语文体特点是清晰、准确、精炼和严密,要求行文简洁、表达客观、内容准确,并且信息量大,强调存在的事实而非某一行为。具体来说,科技文献的特点主要有以下几点。

一、广泛使用被动语态

科技英语的叙述对象通常是客观事物的发展过程和客观存在的真理,往往着眼于演绎论证的结果,而不考虑动作的执行者。第一、二人称使用过多,会造成主观臆断的印象。因此尽量使用第三人称叙述,采用被动语态。据统计,科技英语中的谓语至少三分之一是被动语态。

[例1]　The new ship *was designed* in accordance with the most modern technology and *built* with particularly advanced techniques. The whole process of design *was checked* by computer to ensure that the high strength requirements against underwater explosions *were met*.

分析：该语段由两个句子构成，包含了四个被动语态的句子。

译文：新的舰船是按照最现代化的技术设计，并采用特别先进的方法建造的。整个设计过程由计算机检查，以保证满足防止水下爆炸所需要的高强度要求。

[例2]　Manufacturing process may be classified as unit production, with small quantities being made and mass production with large numbers of identical parts being produced.

译文：制造过程可分为单件生产和批量生产。单件生产就是零件产量小的生产，批量生产就是相同零件产量大的生产。

二、广泛使用非谓语动词

科技文章要求行文简练、结构紧凑，因此往往使用分词短语代替定语从句或状语从句，使用动词不定式短语代替各种从句。

[例1]　A capacitor is a device consisting of two conductors separated by a nonconductor.

分析：本句中的分词短语用作后置定语。

译文：电容器是由被绝缘物体隔开的两个导体组成的一种元件。

[例2]　We keep micrometers in boxes. Our object in doing this is to protect them from rust and dust.

分析：本句可以用不定式短语代替从句来表示目的和功能，即 We keep micrometers in boxes *to protect them from* rust and dust.

译文：将千分尺放在盒子内以防锈和防尘。

[例3]　The cathode ray is a stream of electrons emitted by a heated wire called a cathode.

分析：分词短语作定语。

译文：阴极射线是由称为阴极的受热金属丝发射的电子流。

三、大量使用名词化结构

名词化结构是科技英语的一个很显著的特点，科技英语的作者倾向于将动词或者形容词转化为相应的"名词化"了的动词或形容词来使用，并与介词（如 of）连用，就构成了一个名词化的结构。例如：

[例1]　Television is *the transmission and reception of moving objects image* by radio waves.

译文：电视通过无线电波发射和接收活动物体的图像。

[例2]　*The flow of electrons* is from the negative zinc plate to the positive copper plate.

译文：电子从极性为负的锌板流向正的铜板。

四、长难句较多

科技英语为了适应表达复杂的科技原理和概念的需要，使其表述逻辑严密、结构紧凑，需要大量使用定语从句、状语从句和各种短语,使句子常具有较长的复杂结构。在翻译的时候,要注意搞清各分句的内容和结构,借助语法关系和逻辑关系,使译文前后衔接、相互呼应。

[例1] The classification includes all metal-cutting machinery, the action of which is a progressive cutting away of surplus stock, a gradual reduction in size until the finished dimensions are reached, but excludes sheet metal working machinery and metal-forming and forging machines.

译文:这类机器包括所有的金属切削机床。其作用在于能连续切割多余的坯料,即使坯料尺寸逐渐减小,直到达到规定的值为止。但这类机器不包括金属板加工机、金属成形机与锻压机。

[例2] On account of the accuracy and ease with which resistance measurements may be made and the well-known manner in which resistance varies with temperature, it is common to use this variation to indicate changes in temperature.

译文:电阻的大小是随温度的变化而变化的,而对电阻进行测量既精确又方便,因此,常用电阻的变化来表示温度的变化。

❖ Reading Material

Evolution of FMS

The concept of flexible manufacturing systems was born in London in the 1960s when David Williamson, a research and development engineer, came up with both the name and the concept. At the time he was thinking in terms of a flexible machining system, and it was in a machine shop that the first FMS was installed. His concept was called System 24 because it was scheduled to operate for 24 hours a day under the control of a computer, but otherwise unmanned on the 16-hour night shift. This simple concept of decentralized computer control of machine tools, combined with the idea of using machine tools for 24 hours per day (16 unmanned on night shift), was the beginning of FMSs.

Williamson planned to use NC (numerically controlled) machines to work out a series of machining operations on a wide range of detail parts. Workpieces would be loaded automatically when needed. Each machine would be equipped with a magazine from which tools could be selected systematically to perform a variety of different operations. Included in this overall process were systems for removing chips and cleaning workpieces. This system combined the versatility of computer-controlled machines with very low manning levels.

With the growth in computer-controlled equipment and broader applications developing from metal forming to assembly, the concept of "flexible machining system" was broadened to become what is known today as "flexible manufacturing system", or FMS.

As the first FMS systems were installed in Europe, they followed Williamson's concept, and users quickly discovered that the principles would be ideal for the manufacture of low-volume, high-variety products. The addition of refinements to a FMS to detect and compensate for tool wear were then added to further aid unattended FMS operations. These first FMSs on the market had dual computers: DNC (direct numerical control) for cell control functions and a separate computer to monitor the traffic and management information systems.

Since the 1970s there has been an explosion in system controls and operational enhancements. The programmable controller appeared in the late 1970s, and the personal computer emerged utilizing distributed logic control with many levels of intelligent decision-making capabilities.

Thus, through a conceptual idea originating with David Williamson, it became possible to machine a wide variety of workpieces on few machine with low manning levels productively, reliably, and predictably; this is what FMS is about. In almost any manufacturing industry, FMS will pay dividends as long as it is applied in its broad sense, and not just to define a machine system.

The FMS has evolved rapidly and will continue to evolve because technology continues to evolve, global competition intensifies, and the concept of flexible manufacturing gains wider acceptance. The growth of flexible manufacturing is projected to increase steadily in the years ahead.

In 1984, 56 percent of all FMSs were used for manufacturing machinery and 41 percent for manufacturing transportation components.

Flexible automation is presently feasible for a few machining operation that account for a fraction of the total manufacturing process. However, development efforts continue to expand the FMS's capabilities in the areas of improved diagnostics and sensors, high speed, noncontact, on-line inspection, multifunction or quick spindle head changing machine tools, and extending flexible automation to include forming, heat treating, and assembly. This is why FMS continue to grow and prosper. It feeds on technology evolving and expanding as technology itself evolves and expands.

New Words and Expressions

1. decentralize 使分散
2. enhancement 提高,增加
3. conceptual 概念上的
4. intensify 加强,增强
5. refinement 提炼,精炼

Lesson 12 Mechatronics

Text

In recent years, the discipline known as mechanical engineering has been evolving in response to a growing demand for products of a diversified nature. Among these, the most prevalent examples are those mechanical devices that incorporate some type of electronic circuitry in order to carry out their primary function. The simplest mechanisms, such as gears, pulleys, springs and wheels, have provided the basis for mechanical industry. The electronic technology, on the other hand, is completely a product of the twentieth-century, all of which is created only within the past 100 years. Mechanism plays a vital role in industry. While many industry processes have electronic control systems, it is still mechanisms that deliver the power to do the work. Mechatronics is the term used for the integration of mechanical and electronic engineering disciplines to achieve the objective of developing efficient systems controlled intelligently. The International Federation for the Theory of Machines and Mechanisms defines mechatronics as the synergistic integration of mechanical and electronic engineering, electronic controls, and software in the design of products and manufacturing processes as indicated in Figure 12.1. Mechatronics is much more significant than its implying of combined words. It is changing the way by which we design and produce the next generation of high technology products.

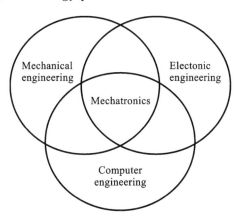

Figure 12.1 Mechatronics integrates mechanical, electrical and computer engineering

Mechatronics is an interdisciplinary technical discipline, built upon the basis of classical mechanical, electrical and electronic engineering, binding these sciences not only with one another, but also with computer science and software engineering. The central focus of mechatronics is the integral development of systems from technical components ("mecha"), which are to be intelligently controlled ("tronic"). Thus a system composed of mechanical and electrical parts, overlaid with sensors which record information, microprocessors which interpret, process and analyze the information and assemblies which then react upon this information, becomes a complete mechatronic system as shown in Figure 12.2.

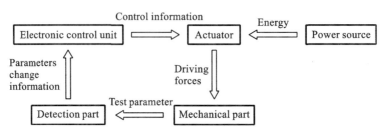

Figure 12.2 Basic elements of a mechatronics system

Modern industrial systems and components typically feature various sensors, actuators, and controllers integrated into complex configurations that incorporate skills from various engineering disciplines. To design and service this equipment, global companies often use engineering teams familiar with mechatronic system technologies. Some of the key technical skills include mechanical, electrical, computer, and industrial engineering as well as control systems, computer simulation, robotics, and human factors. For example, different types of conveyor and robotic elements may be applied to transport materials, assemble components, and then move the finished goods within a manufacturing facility.

The basis of mechatronic system design dictates that the mechatronic system design processes include the simultaneous integration of the cross-discipline components and systems. This integrated / concurrent approach toward system design is sought in an effort to yield performance which is simply unattainable via the traditional discipline isolated design methods. The isolated approach prevents the insights apparent from the concurrent interplay between system components. One of the major problems in the design of mechatronic system is the integration of hardware and software components and formal communication among the various technical disciplines involved in the design of the mechatronic application. As the size and complexity of mechatronic system's increases i.e. large-scale, the system design goes beyond hardware and algorithms. Mechatronic has dictated the need for a fundamental change in the hardware and software design process in order to ensure affordability, reliability, maintainability, adaptability and a built-in growth potential for large-scale applications. One of the major challenges facing the mechatronic community in the development of next generation smart systems, lies within the hardware and software integration requirements of multidiscipline design inherent in these mechatronic systems. Design and integration of these large scale systems need to start at a higher level then we are used to.

New Words and Expressions

1. mechatronics   机电一体化；机械电子学   2. electronic engineering   电子工程

3. integration　集成；综合
4. interdisciplinary　跨学科的，跨领域的
5. mechanical engineering　机械工程
6. actuator　制动器，执行器
7. controller　控制器
8. reliability　可靠性
9. encompass　包含；包围，环绕；完成
10. maintainability　可维护性，可维修性
11. artificial intelligence　人工智能
12. adaptability　适应性，可变性
13. affordability　支付能力，可购性
14. built-in　嵌入的，内置的

Notes

(1) The International Federation for the Theory of Machines and Mechanisms defines mechatronics as the synergistic integration of mechanical and electronic engineering, electronic controls, and software in the design of products and manufacturing processes as indicated in Figure 12.1.

国际机器理论与机构学联合会将机电一体化定义为在产品设计和制造过程中，机械工程、电子工程、电子控制以及与产品设计和制造过程相关的计算机软件的有机结合，如图12.1所示。

(2) One of the major problems in the design of mechatronic system is the integration of hardware and software components and formal communication among the various technical disciplines involved in the design of the mechatronic application.

机电一体化系统设计的一个主要问题是软、硬件部分的集成以及机电一体化应用产品设计中所涉及的各种技术学科之间的相互交叉及有机整合。

(3) One of the major challenges facing the mechatronic community in the development of next generation smart systems, lies within the hardware and software integration requirements of multidiscipline design inherent in these mechatronic systems.

在下一代智能系统的开发过程中机电一体化技术面临的重大挑战之一，就在于机电系统的多学科综合的特性对硬件和软件的集成提出了进一步的要求。

❖ Translation Skills

英汉语序比较与翻译语序处理

英语和汉语语序特征的不同点主要是两种语言中定语和状语的位置不同。汉语语序相对稳定，而英语语序较为灵活。

一、定语的语序

英语中单个代词、形容词、数词、分词、动名词及所有格名词等作定语时，往往置于被修饰词之前，与汉语语序基本相同，因此汉译时可照译不变。

[例1]　The law is known as the Second Law of Newton.

译文：这个定律称为牛顿第二定律。

[例2]　Figure 3 shows the motion of a moving object.

译文：图3显示了一个运动物体的运动情况。

分词或形容词作定语时，有时后置，此时分词往往有较强的动词意义。翻译时可前置或后置，或译成其他成分。

[例3] The greater the force applied, the greater the acceleration.

译文：外（加的）力越大，加速度就越大。

[例4] The assumptions made will affect the type of series obtained.

译文：所作的假设会影响获得的级数的类型。

二、短语定语的位置

英语中定语短语有分词短语、动词不定式短语、介词短语及形容词短语等。一般来说，这些定语均置于被修饰的中心词之后，汉译时一般仍需前置。定语语序则可按前述规律安排。

[例1] Some substances having very high resistance are called insulators.

译文：电阻很高的物质称为绝缘体。

[例2] A thermometer is an instrument to show the temperature of the air or other surroundings.

译文：温度计是指示气温或其他环境温度的仪器。

"表示方式或程度的副词＋现在分词或过去分词"形式的短语，往往置于所修饰的名词之前。汉译时可不改变词序。

[例3] Laser-guided shells give conventional artillery a greatly improved anti-tank capability.

译文：激光制导炮弹大大改善了常规火炮的反坦克能力。

若句中有两个短语作定语并修饰句中同一个词，一般是较短的或关系较近的放在前面，较长的或关系较远的放在后面。汉译时可先译较近的，再译较远的，均译为前置定语。

[例4] The ability of a metal to be drawn into a wire is known as ductility.

译文：金属被拉成金属丝的性能称为延性。

英语中定语短语有时会被句子中的谓语或状语分隔。之所以会产生这种现象，有两种原因：其一是主语的定语长而谓语短，这时可将主语的定语放到谓语之后，以使句子平衡；其二是状语短而定语长，形成分隔。若不注意这种分隔现象，往往会造成误译。

[例5] Temperature plays the same part in the flow of heat that pressure does in the flow of fluids.

分析：that pressure…是 part 的定语。

译文：温度在热的传导体中所起的作用和压力在液体的流动中所起的作用相同。

[例6] Many form of apparatus have been devised by which a more accurate knowledge of blood pressure can be obtained.

分析：by which…是 apparatus 的定语。

译文：人们已经设计出许多医疗器械，通过这些医疗器械，可以对血压有更为准确的了解。

英语含有数量概念的名词或表示计量单位的名词与介词 of 构成的短语定语，通常置于所修饰名词之前。这恰好与汉语的语序相同，汉译时可不必变动语序。

[例7] When a wire is broken by bending it back and forth rapidly, some of the work is transformed into heat and the wire gets out.

译文：当把导线快速来回弯曲折断时，部分功就转换成热，所以导线变热。

[例8] High-performance tourmaline plate releases a large of negative ions.
译文：高性能电石板释放大量负离子。

三、状语的语序

英语副词可修饰形容词、数词、介词短语、代词或另一副词等。这时，副词往往位于所修饰的词之前。翻译时一般不需要改变词序。

[例1] The boats travel very slowly when the wind is foul.
分析：副词 very 修饰另一个副词 slowly。
译文：在逆风的时候，船走得很慢。

[例2] The attractive force between the molecules is negligibly small.
分析：副词 negligibly 修饰形容词 small。
译文：分子间的吸引力小得可以忽略不计。

表示程度及方式的副词修饰动词时，其位置可在动词之前或是之后。但修饰不及物动词时常在其后。汉译时，一般置于动词之前，但有时也可置于其后（特别是副词的比较级或最高级作状语时），可在动词之后加"得"或"很"等词。

[例3] Electronic computers can work very quickly and with make no errors.
译文：电子计算机能非常快地工作，并且不出错。

[例4] The temperature of copper increases more rapidly than that of water.
译文：铜温比水温增加得快。

表示频度及不确定时间概念的副词（例如 always, never, already, often, seldom, quite, almost, hardly, usually, generally…）作状语，通常置于行为动词之前。如有系词、情态动词、助动词，则置于这些词之后。汉译时，通常译在主要动词之前。

[例5] A microcomputer generally used by only person at a time uses a microprocessor chip as its CPU.
译文：一台微型计算机在一段时间里通常仅由一个人使用。它用一个微处理器芯片作为 CPU。

修饰全句的副词状语一般置于句首，汉译时可不改变词序。

[例6] Obviously the physical condition of the cosmonaut is of immense important.
译文：很显然，宇宙飞行员的身体条件是极其重要的。

[例7] Actually, the whole mass of the atom is concentrated in the nucleus.
译文：实际上，原子的整个质量都集中在原子核里。

四、短语状语的位置

由介词短语、分词短语及动词不定式短语构成的状语，可置于被修饰动词之前或之后。汉译时酌情前置或后置。

[例1] In order to get different speeds many lathes are equipped with multi-speed gear-boxes.
译文：为得到不同速度，许多车床装有多级变速齿轮箱。

[例2] A resistor is placed in parallel with another only to make the current become great.
译文：将一只电阻与另一只电阻并联起来，电流反而变大。

英语中如果同时有几个状语，一般是较短的在前、较长的在后，地点状语在时间状语之前。

汉译时需要调整词序。

[例3] In a conductor, the free electrons can be made to move freely from atom to atom under a definite potential force.

分析：原文的状语顺序按由短到长的顺序排列，汉译时倒换顺序。

译文：在一定的电势力的作用下，导电体中的自由电子能在原子间自由移动。

英语中的时间状语和地点状语，一般是单位小的、范围窄的在前，单位大的、范围宽的在后。汉译时应颠倒次序。

[例4] China successfully launched its first man-made earth satellite in April 1970.

译文：中国于1970年4月成功发射了第一颗人造地球卫星。

❖ Reading Material

Mechatronic System Design

One of the most challenging problems in mechatronics systems design is the development of system architecture, e. g. , selection of hardware (actuators, sensors, devices, power electronics, ICs, microcontrollers, and DSPs) and software (environment and computation algorithms to perform sensing and control, information flow and data acquisition, simulation, visualization, and virtual prototyping). Attempts to design state-of-the-art man-made mechatronics systems and to guarantee the integrated design can be pursued through analysis of complex patterns and paradigms of evolutionary developed biological systems. Recent trends in engineering have increased the emphasis on integrated analysis, design and control of advanced mechanical systems. The scope of mechantronics system has continued to expand, and in addition to actuators, sensors, power electronics, ICs microprocessors, DSPs, as well as input/output devices, many other subsystems must be integrated, even through the basic foundations have been developed, some urgent areas have been downgraded, less emphasized and covered. The basis of mechatronic system design dictates that the mechatronic system design processes include the simultaneous integration of the cross-discipline components and systems. This integrated / concurrent approach toward system design is sought in an effort to yield performance which is simply unattainable via the traditional discipline isolated design methods. The isolated approach prevents the insights apparent from the concurrent interplay between system components. This is due to the fact that poor communication and differing design strategies prevail when the elements of the system are not considered as a whole at design time. Ideally, the mechatronic system design process offers practitioners from the various disciplines a unified and standardized means of expressing and considering the flow of energy and information between the multi-domain or multidisciplinary components of the system.

The mechatronic design process benefits from a formal consideration of information and energy transfer between components within the given system. A generalized diagram of this energy and information transfer process is shown in Figure 12. 3. With this consideration comes the substantial task of representing and analyzing the various system components in a level of detail which captures a prescribed level of resolution with regard to the response characteristics of the components and system as a whole. Only when this task is achieved successfully will the results reflect the benefit of the performance, reliability and optimality offered by the mechatronic approach.

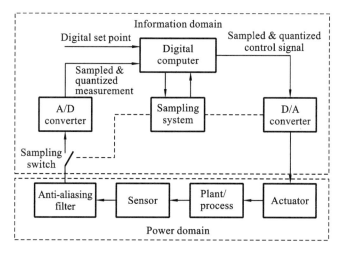

Figure 12.3 Flow of energy and information between domains

New Words and Expressions

1. microcontroller   微控制器
2. prototyping   样机研究;原型设计
3. state-of-the-art   最先进的,已经发展的
4. concurrent   并发的,同时的
5. is sought in   寻求
6. multi-domain   多领域
7. multidisciplinary   多学科的
8. resolution   分辨率,决心
9. with regard to   关于,就……而论

Lesson 13   Hydraulic and Pneumatic Systems

Text

1. Hydraulic systems

A hydraulic system is a drive or transmission system that uses pressurized hydraulic fluid to drive hydraulic machinery. A hydraulic system is developed when the various components such as reservoir, pump, valve, and motor are combined and joined together with the necessary pipe, tubing, and fittings (see Figure 13.1).

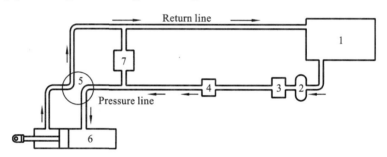

Figure 13.1   Schematic diagram of a hydraulic system
1—reservoir; 2—pump; 3—filter; 4—pressure switch; 5—selector valve; 6—actuator; 7—relief valve

The pump (3) starts turning and draws fluid from the reservoir (1) and pushes it out into the system through the selector valve (5). Since fluid is incompressible, pressure will start to build and the actuator (6) will start to move immediately to the left. The fluid leaving the actuator (6) will go right back through the selector valve (5) to the reservoir(1). The fluid entering the pump (3) also goes through a filter (2) to keep debris out of the system. The actuator (6) is moving a heavy object. The pressure in the system will only be as high as it takes to move the object. When the pressure in the system excesses the settings, the pressure switch (4) triggers and pressure relief valve (7) function and bypass fluid back to the reservoir, to prevent damage to the system.

1) Hydraulic pumps

The purpose of a hydraulic pump is to pressurize the hydraulic fluid so that work may be performed. Every hydraulic system uses one or more pumps to pressurize the hydraulic fluid. The fluid under pressure, in turn, performs work in the output section of the hydraulic system. Thus the pressures fluid may be used to move a piston in a cylinder or to turn the shaft of a hydraulic motor. Common types of hydraulic pumps to hydraulic machinery applications are gear pumps, vane pumps, and piston pumps, etc(see Figure 13.2).

A gear pump uses the meshing of gears to pump fluid by displacement. They are one of

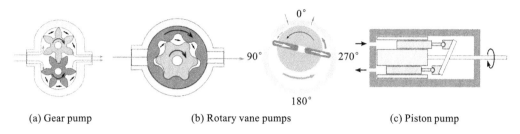

(a) Gear pump　　　　　(b) Rotary vane pumps　　　　(c) Piston pump

Figure 13.2　Various hydraulic pump types

the most common types of pumps for hydraulic fluid power applications. There are two main variations namely external gear pumps which use two external spur gears, and internal gear pumps which use an external and an internal spur gear. Gear pumps are positive displacement, meaning they pump a constant amount of fluid for each revolution. Some gear pumps are designed to function as either a motor or a pump.

A rotary vane pump is a positive-displacement pump that consists of vanes mounted to a rotor that rotates inside of a cavity. In some cases these vanes can be variable length and/or tensioned to maintain contact with the walls as the pump rotates. On the intake side of the pump, the vane chambers are increasing in volume. These increasing volume vane chambers are filled with fluid forced in by the inlet pressure. Inlet pressure is actually the pressure from the system being pumped, often just the atmosphere. On the discharge side of the pump, the vane chambers are decreasing in volume, forcing fluid out of the pump.

Piston pumps are more expensive than gear or vane pumps, but provide longer life operating at higher pressure, with difficult fluids and longer continuous duty cycles. An axial piston pump has a number (usually an odd number) of pistons arranged in a circular array within a housing which is commonly referred to as a cylinder block, rotor or barrel. As the piston moves toward the valve plate, fluid is pushed or displaced through the discharge port of the valve plate. A radial piston pump is normally used for very high pressure at small flows. In a radial piston pump, the working pistons extend in a radial direction symmetrically around the drive shaft. The piston starts in the inner dead center (IDC) with suction process. After a rotation angle of 180° it is finished and the workspace of the piston is filled with the fluid. The piston is now in the outer dead center (ODC). From this point on the piston displaces the previously sucked medium in the pressure channel of the pump.

2) Control valves

Valves are used to control and regulate the flow direction of the hydraulic fluid, the pressure, the flow rate, and consequently, the flow velocity. There are four types of control valves selected in accordance with the problem in question.

Directional control valves　Directional control valves control the direction of flow of the hydraulic fluid and, thus the direction of motion and the position of the working components. They usually consist of a spool inside a cylinder which is mechanically or electrically controlled. The movement of the spool restricts or permits the flow, thus it controls the fluid flow.

Pressure control valves Pressure control valves are used in hydraulic circuits to maintain desired pressure levels in various parts of the circuits. A pressure control valve maintains the pressure level by diverting higher pressure fluid to a lower pressure area, or restricting flow into another area. Valves that divert fluid can be safety, relief, counterbalance, sequence, and unloading types. Valves that restrict flow into another area can be of the reducing type. Relief, sequence, unloading and counterbalance valves are normally closed, that are partially or fully opened while performing their design function. A reducing valve is normally opened valve that restricts and finally blocks the fluid flow into a secondarily area.

Flow control valves These valves interact with pressure valves to affect flow rate. They make it possible to control or regulate the speed of motion of the power components, where the flow rate is constant, division of flow must take place. This is generally affected through the interaction of the flow control valve with a pressure valve.

Check valves A check valve or non-return valve is a mechanical device which normally allows fluid to flow through it in only one direction. There are various types of check valves used in a wide variety of applications. Although they are available in a wide range of sizes and costs, check valves generally are very small, simple, and/or inexpensive. Check valves work automatically and most are not controlled by a person or any external control; accordingly, most do not have any valve handle or stem.

3) Advantages of hydraulic systems

Variable speed and reversible The actuator (linear or rotary) of a hydraulic system can be driven from maximum speeds to reduced speeds by varying the pump delivery using a flow control valve. In all cases power is continuously transmitted whilst speed changes take place. In addition, a hydraulic actuator can be reversed instantly while in full motion without damage.

Higher loads and small packages The hydraulic working fluid is basically incompressible, leading to a minimum of spring action. The fluid in a hydraulic circuit is capable of moving much higher loads and providing much higher forces due to the incompressibility. Hydraulic components, because of there high speed and pressure capabilities, can provide high power output with very small weigh and size.

Overload protection Pressure relief valves in a hydraulic system protect it from overload damage. When the load exceeds the valve setting, pump delivery is directed to the tank. The result is definite limit to torque or force output. Pressure relief valve also provide a means of setting a machine for a specified amount of torque or force, as in a chucking or clamping operation.

2. Pneumatic systems

Pneumatics is a branch of technology that deals with the study and application of pressurized gas to effect mechanical motion. A pneumatic circuit comprises the following components: a compressor, pneumatic cylinders, air tank, valves, pneumatic hoses, etc. Figure 13.3 shows a pneumatic circuit with air-saving regulator. A 5-way spool valve, piped with a dual-pressure inlet as shown, can give normal cycle time while conserving plant compressed air. Return pressure is set on the regulator supplying the cylinder rod end at the lowest possible pressure that maintains cycle integrity. A reduction as small as 20 psi below working pressure can pay for the regulator in a short time. Shifting the 5-way valve starts the cylinder extending. There will be a brief lunge as the lower-pressure air in the rod end compresses to hold back against the higher pressure in the cap end. To control cycle time, adjust cylinder speed with the rod-end meter-out flow control. When the 5-way valve shifts again to return the cylinder, the meter-out flow control on the cap end must be adjusted for a faster rate because return power is limited.

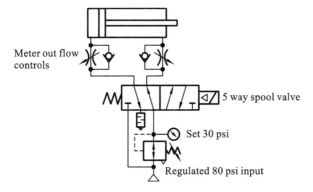

Figure 13.3　Schematic diagram of a pneumatic system

Pneumatic control systems are widely used in our society, especially in the industrial sectors for the driving of automatic machines. Advantages of pneumatic systems include:

1) Simple in design and control

The designs of pneumatic components are relatively simple. They are thus more suitable for use in simple automatic control systems. The speeds of rectilinear and oscillating movement of pneumatic systems are easy to adjust and subject to few limitations. The pressure and the volume of air can easily be adjusted by a pressure regulator. As pneumatic components are not expensive, the costs of pneumatic systems are quite low. Moreover, as pneumatic systems are very durable, the cost of repair is significantly lower than that of other systems.

2) Reliability and safety

Many factories have equipped their production lines with compressed air supplies and movable compressors. Moreover, the use of compressed air is not restricted by distance, as it can easily be transported through pipes. Pneumatic systems are safer than electromotive systems because they can work in inflammable environment without causing fire or explosion. Apart from that, overloading in pneumatic system will only lead to sliding or cessation of operation.

3) Adaptability to harsh environment

Compared to the elements of other systems, compressed air is less affected by high temperature, dust, corrosion, etc. The operation of pneumatic systems does not produce pollutants. Therefore, pneumatic systems can work in environments that demand high level of cleanliness. One example is the production lines of integrated circuits.

Both pneumatics and hydraulics are applications of fluid power. Pneumatics uses an easily compressible gas such as air or a suitable pure gas, while hydraulics uses relatively incompressible liquid media such as oil. Most industrial pneumatic applications use pressures of about 80 to 100 psi (550 to 690 kPa). Hydraulics applications commonly use from 1,000 to 5,000 psi (6.9 to 34 MPa), but specialized applications may exceed 10,000 psi (69 MPa).

New Words and Expressions

1. hydraulic   液压的,液力的
2. spool   阀杆,线轴
3. pneumatic   充气的,气动的,气压的
4. check valve   单向阀,止回阀
5. tubing   (金属、塑料等的)管材
6. compressor   压缩机,压气机
7. fitting   连接件,接头
8. cylinder   汽缸,圆筒
9. reservoir   油箱,蓄水池,储液器
10. hose   软管,胶皮管
11. actuator   执行机构,促动器
12. spool valve   滑阀,柱式阀
13. trigger   触发;触发器,启动装置
14. meter-out flow control   出口节流控制
15. chamber   腔室,室
16. cap end   活塞柱头
17. displacement   排量,位移
18. rectilinear   直线的,形成直线的
19. discharge   卸料;流出,排出
20. hold back   阻碍,抑制

Notes

(1) Piston pumps are more expensive than gear or vane pumps, but provide longer life operating at higher pressure, with difficult fluids and longer continuous duty cycles.

柱塞式液压泵比齿轮泵和叶片泵造价高,适用于高压工作环境和泵送特性复杂的流体,性能可靠、使用寿命长。

(2) Valves are used to control and regulate the flow direction of the hydraulic fluid, the pressure, the flow rate, and consequently, the flow velocity.

液压阀用来调节液压管路中的液体流向、工作压力和流量,进而调节流速。

(3) Relief, sequence, unloading and counterbalance valves are normally closed, that are partially or fully opened while performing their design function. A reducing valve is normally opened valve that restricts and finally blocks the fluid flow into a secondarily area.

溢流阀、顺序阀、卸荷阀和平衡阀属于常闭阀,根据预设功能要求,这些阀在工作时处于全开或部分开启状态;减压阀一般为常开阀,用来限制和阻断液压油进入下游工作区。

(4) There will be a brief lunge as the lower-pressure air in the rod end compresses to hold back against the higher pressure in the cap end.

由于处于活塞杆一端的低压空气受到压缩,会阻碍活塞柱头一端的高压空气进入,因此吸气过程中会存在短暂的冲击。

(5) When the 5-way valve shifts again to return the cylinder, the meter-out flow control

on the cap end must be adjusted for a faster rate because return power is limited.

当改变 5 通滑阀的流向,使汽缸活塞返程时,须将与活塞环一端相连的排气流量控制阀的流速调整到最大值,因为返程动力的供给是有限的。

❖ Writing Training

科技术语定义的表达

定义是用来对某种事物的本质特征或某个概念的内涵及外延进行确切而简要的说明的一种文体表达方式。通过定义叙述,可使比较模糊、抽象的科技术语或概念更易于理解,使信息交流者对事物的理解有一个共同的基础。

(一) 归纳式定义

在科技术语定义写作中,要求定义表达精确、全面。逻辑归类法是定义的常用写作方法。当需要说明某种事物的特性、种类或功能时,即主题句为回答诸如某种事物"是什么""是什么类型"或"有什么功能"等情况时,往往用下面的逻辑句型表达:

S(所要定义的术语)+V+O(属)+wh…/ that 定语从句(事物的专属特性,种差)。

这种定义的表达形式常出现在词典、百科全书、科技著作或科技论文中。

[例1] An *microscope* is an *optical instrument* which *consists of a lens or combination of lenses for making enlarged images of minute objects*.

译文:显微镜是一种由一个或一组透镜组成、用来对微小物体进行放大的光学仪器。

[例2] *Gravity* is *the attraction between objects in the universe* that *keeps our moon going around the earth*, that *keeps the earth and the other planets going around the sun*, and that *brings down to the earth an object tossing in the air*.

译文:重力是宇宙中物体之间的一种引力,是使月球绕着地球转动,地球和其他星球绕着太阳转动以及抛向空中的物体落向地面的力。

写作归类式定义时,一定要抓住所要定义的事物的本质。选好事物的"属"非常重要。属概念的范围要尽量窄,否则"种差"的内容会非常庞杂。"种差"一定要揭示它与同一属概念中其他种之间的差异。有时用一个简单句给一个术语下定义也能表达清楚。

[例3] The deadweight of a vessel is defined as weight of cargo plus fuel and consumable stores.

译文:一艘船舶所载运货物的重量与船载燃油和消耗品的重量之和称为总载重量。

(二) 广延式定义

在某些情况下,为了保证所下定义更加清楚、准确,消除可能产生的模糊认识,有时需要事物进行更加详细的说明。广延式定义表达就是运用各种方法(如举例、对比、列出组成、排除等)对事物本质特点的内涵和外延作进一步说明。广延式定义一般很长,甚至形成整篇文章。

[例1] Physics is the science that deals with those phonemes of matter involving no change of chemical composition. *It includes the science of matter and motion, mechanics, heat, light, sound, electricity, and the branches of science devoted to radiation and atomic structure.*

译文:物理学是研究物质(化学组成不变)的内在规律及现象的一门科学,内容涉及物质结

构、运动学、力学、热学、光学、电学以及放射性和原子结构科学等。（列出组成）

［例2］ An unemployed person is a member of the working population who does not have a job. *For example, in UK, the working population includes all the people who are*：(1) *physically and mentally capable of work*；(2) *actively looking for a job*；and(3) *between 16 to 65 for men and between 16 to 60 for women*.

译文：失业人员是指具有劳动能力而没有找到工作的人员。例如在英国，具备以下条件的人员属于失业人员：(1) 身心健康，能够从事某种工作；(2) 在主动寻找工作；(3) 男性年龄为16～65岁，女性年龄在16～60岁之间。（举例）

［例3］ An atom is *similar to a biological cell of a microscope model of a solar system*. It is the smallest unit of chemical element, composed of a central nucleus surrounded by electronically charged particles.

译文：原子与在显微镜下观察到的太阳系中的生物体细胞相似，由原子核和核外电子组成，是构成物质的最小单元。

［例4］ Unemployment rate is the percentage of the labor force officially jobless. *When calculating the level of unemployment, a government usually only counts those people who register as unemployment and claim benefit. Those people seeking jobs that either do not register or do not claim benefit are usually excluded from the official figures*.

译文：失业率是指官方统计的失业人数占总劳动人口的比例。政府在计算失业人口时，一般只统计已注册失业并且申请失业救济金的人员，而对于那些正在寻找工作但没有注册失业和申请救济金的人员，一般不计入其中。（排除法）

❖ Reading Material

Components of Hydraulic Systems and Application

Other basic components of hydraulic systems

Hydraulic cylinders：A hydraulic cylinder is a mechanical actuator that is used to give a linear force through a linear stroke. Hydraulic cylinders are broadly classified into two categories: single acting cylinders and double acting cylinders. Single acting cylinder applies force only in one direction. The return motion is accomplished by releasing the pressure when the piston is moved back to its original position by a spring or some external force. Double acting cylinder is operated by hydraulic fluid in both directions and it is capable of a power stroke either way. In such cylinders, fluid ports are fitted to each end, to function alternately as inlet and outlet ports. The maximum output force available is slightly less than that obtainable from a single acting cylinder due to backpressure generated in the return line. Further in the reverse direction piston rod seals may offer frictional resistance.

Hydraulic motors：Hydraulic motors and pumps are essentially similar in construction and most of the pumps can act as motors. Hydraulic motor can act as a direct alternative to the electric motor with built in infinitely gearbox. Even clutch is unnecessary with the transmission driven by hydraulic motor. Thus no clutch gear changing is involved and the hydraulic motor can remain stalled indefinitely without harm. Even frequent reversing does not cause damage and it can be used for dynamic braking. Like hydraulic pump, motors can also be classified as fixed type and variable delivery type. According to the constructional feature hydraulic motors are termed as gear type, vane type, piston type, etc.

Reservoirs: The hydraulic fluid reservoir holds excess hydraulic fluid to accommodate volume changes from: cylinder extension and contraction, temperature driven expansion and contraction, and leaks. The reservoir is also designed to aid in separation of air from the fluid and also work as a heat accumulator to cover losses in the system when peak power is used. Design engineers are always pressured to reduce the size of hydraulic reservoirs, while equipment operators always appreciate larger reservoirs. Reservoirs can also help separate dirt and other particulate from the oil, as the particulate will generally settle to the bottom of the tank.

Hydraulic fluids: Hydraulic fluid is the life of the hydraulic circuit. It is usually petroleum oil with various additives. Some hydraulic machines require fire resistant fluids, depending on their applications. In some factories where food is prepared, either an edible oil or water is used as a working fluid for health and safety reasons. In addition to transferring energy, hydraulic fluid needs to lubricate components, suspend contaminants and metal filings for transport to the filter, and to function well to several hundred degrees Fahrenheit or Celsius.

Accumulators: Accumulators are a common part of hydraulic machinery. Their function is to store energy by using pressurized gas. One type is a tube with a floating piston. On one side of the piston is a charge of pressurized gas and on the other side is the fluid. Bladders are used in other designs.

Filters: Filters are an important part of hydraulic systems. Metal particles are continually produced by mechanical components and need to be removed along with other contaminants. Filters may be positioned in many locations. The filter may be located between the reservoir and the pump intake. Blockage of the filter will cause cavitation and possibly failure of the pump. Sometimes the filter is located between the pump and the control valves. This arrangement is more expensive, since the filter housing is pressurized, but eliminates cavitations' problems and protects the control valve from pump failures. The third common filter location is just before the return line enters the reservoir. This location is relatively insensitive to blockage and does not require a pressurized housing, but contaminants that enter the reservoir from external sources are not filtered until passing through the system at least once.

Tubes, pipes and hoses: Hydraulic tubes are seamless steel precision pipes, specially manufactured for hydraulics. The tubes have standard sizes for different pressure ranges, with standard diameters up to 100 mm. The tubes are supplied by manufacturers in lengths of 6 m, cleaned, oiled and plugged. The tubes are interconnected by different types of flanges (especially for the larger sizes and pressures), welding cones/nipples (with O-ring seal), and several types of flare connection and by cut-rings. Direct joining of tubes by welding is not acceptable since the interior cannot be inspected. Hydraulic pipes are generally used for low pressure and/or larger sizes. They can be connected by threaded connections, but usually by welds. Because of the larger diameters the pipe can usually be inspected internally after welding. Hydraulic hose is graded by pressure, temperature, and fluid compatibility. Hoses are used usually to provide flexibility for machine operation or maintenance. The hose is built up with rubber and steel layers. The weakest part of the high pressure hose is the connection of the hose to the fitting. Another disadvantage of hoses is the shorter life of rubber which requires periodic replacement, usually at five to seven year intervals.

Application of hydraulic power to machine tools

The application of hydraulic power to machine tools goes back to the mid-1930s when CIP, a Swiss manufacturer of Jig Boring machinery incorporated hydraulics to move the tool bed. Since then, Hydraulic power has become an integral part of production machines that perform pressing, drilling, cylinder boring and honing, crankshaft and camshaft grinding, gear milling, and other automobile production jobs.

Hydraulic machine tool drive offers a great many advantages. One of them is that it can give infinitely-variable speed control over wide ranges. In addition it can change the direction of drive as easily as it can vary the speed. As in many other types of machine, many complex mechanical linkages can be simplified or even wholly eliminated by the used of hydraulics. The flexibility and resilience of hydraulic power is another great virtue of this form of drive. Apart from the smoothness of operation thus obtained, a great improvement is usually found in the surface finish on the work and the tool can make heavier cuts without detriment and will last considerably longer with regrinding.

High volume production utilizes multi-station indexing machines where multiple hydraulic machining heads are arranged in clusters to perform multiple operations on parts simultaneously and from both sides. The work pieces are clamped, moved under successive heads and then unloaded in a matter of seconds. Hydraulic power can supply any sort of rotation or linear force needed in machine tools design. In terms of rotation, such as tool heads and spindle drives, hydraulic motors come in a wide range of sizes and power. They have a size to power ratio that is superior to most other rotary motion drives and are generally much less noisy. A motor the size of a soft-drink can is capable of producing up to 5 hp(1 hp = 735 W). If linear motion is required, such as advancing a tool head, hydraulic pistons can be used or motor-driven screw or rack and pinion drives. Hydraulic chucks are available in many sizes to hold the work piece or feed bar stock in an automatic turning operation. Hydraulic pistons provide the steady power strokes in broaching operations and surface grinding. The beauty of hydraulic power is its adaptability within the structure of the machining center. Small tool heads or spindles can be located within the machining centers where ever needed, the only connections being a pair of small hoses and electrical wires for the sensors and controls.

New Words and Expressions

1. stroke 冲程
2. sensor 传感器
3. reservoir 油箱
4. hydraulic cylinder 液压缸
5. accumulator 蓄能器
6. hydraulic motor 液压马达
7. bladder 囊,袋,囊状物
8. machine tool 机床
9. cavitation 空洞形成,气穴现象
10. infinitely-variable speed 无级变速
11. nipple 乳头状接头,加油嘴
12. machining center 加工中心
13. broaching 扩孔,铰孔
14. cut-ring 胀圈

Lesson 14 Modern Design Theories and Methods

Text

Modern design theories and methods refer to the emerging design theories and methods deviating from those traditional ones which focus on strength and stiffness. They are featured by, for example, using computer to carry out optimal design, computer aided design, reliability calculation etc. Therefore, a variety of topics and include all non-traditional emerging design theories and methods are covered. Usually the following topics are included.

1. Optimal Design

Optimal design is a design method to select the best solution from many a scheme. Optimal design is based on optimization theory and computer calculation to determine the design variables according to the problem, establish the objective functions and find an optimal scheme which satisfies all given constrained conditions. The optimal technology was first put into use militarily in World War II. Mechanical optimal design started in 1960s and was employed in production widely. In 1967, R. L. Fox and K. D. Willmert published their results on curve generating linkage using optimum design. Thereafter, C. S. Beightler et al. solved optimum design of a hydrodynamic journal bearing by geometric programming. Mechanical optimal problems usually belong to non-linear programming scope, with the development of mathematical theory and computer technology, the optimal design has been widely applied in engineering and become an independent engineering discipline.

The mathematical model of an optimal design problem is composed of three elements: design variables, objective function and constraint conditions. It is expressed as: to find a set of design variable which can satisfy all the constraint conditions and optimize the objective function. The mathematical model of a common constraint optimal problem reads

$$\begin{cases} \min f(x), & x \in D \subset \mathbf{R}^n \\ \text{s.t.} \quad h_v(x)=0, & v=1,2,\cdots,p \\ \quad g_u(x) \leqslant 0, & u=1,2,\cdots,m \end{cases}$$

2. Tribology Design

Tribology design plays an important role in energy saving, emission reduction and environmental protection. It is a key part in designing mechanical products, because tribology problems widely exist in dynamical systems, transmission system and mechanical processing procedures, etc. Tribology design is helpful in enhancement of creative design and performance of products and can render effective measures and solutions to major technical problems in engineering.

In mechanical design, the effect of Tribology design are mainly embodied in two aspects: one is to reduce excessive friction and wear in the process; the other is to transmit motion and force by friction. Tribology is the science and technology of two interacting bodies in relative motion and in which, friction, wear and lubrication are studied. The word was coined in 1966 based on the Greek word "tribos", which means rubbing. It is not only closely related to our daily life, such as face washing or teeth brushing, but also technologies and sciences in numerous fields, for example, astronautics (space-tribology) and seismology (geo-tribology). In its broad scope, it is an interdisciplinary subject embracing mathematics, chemistry, physics, material sciences, mechanical engineering and so on. The task of Tribology design is based on mechanics, material science and surface science to investigate the novel principles and functions of new product design and finalize low friction and high wear resistance, in order to achieve aims such as energy saving, long life and higher working performance.

According to the theory of Tribology design, contact conditions between machine elements can be divided into dry contact, boundary contact, mixed contact, thin film lubrication, elastohydrodynamic lubrication (EHL), hydrodynamic lubrication and hydrostatic lubrication, as shown in Figure 14.1. In Figure 14.1, the mixed contact is a state where several contact conditions coexist. Table 14.1 presents the typical film thickness, formation mechanism and application of the above contact conditions.

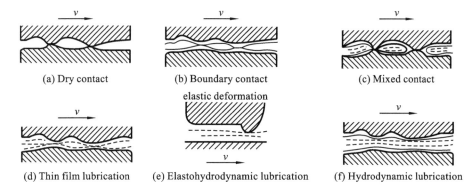

Figure 14.1 Contact conditions

Table 14.1 Features of the contact conditions

Contact conditions	Film thickness	Mechanism of oil formation	application
Dry contact	1~10 nm	Oxidation layer, gas adsorbed layer	Tribological pair with no lubrication or self-lubrication
Boundary contact	1~50 nm	Lubrication film formed by lubricant molecule reacted with the metal surface by physical or chemical reaction	High precision tribological pair under low-speed and heavy load

continue

Contact conditions	Film thickness	Mechanism of oil formation	application
Thin film lubrication	10～100 nm	Formed by the dynamic effect produced by two interacting bodies in relative motion	High precision line/point contact tribological pair under low-speed, ex, high precision rolling bearings
Elastohydrodynamic lubrication	0.1～1 μm	The same as above	High precision line/point contact tribological pair under middle and high speed, ex, gears or rolling bearings
Hydrodynamic lubrication	1～100 μm	The same as above	Face contact tribological pair under middle and high speed, ex, sliding bearings
Hydrostatic lubrication	1～100 μm	Formed by external pressure to deliver lubricant into gap between the contact surfaces	Face contact tribological pair under low speed, ex, sliding bearings or guides

3. Computer aided design

Computer-aided design (CAD), also known as computer-aided design and drafting (CADD), is the use of computer systems to assist in the creation, modification, analysis, or optimization of a design. Computer-aided drafting describes the process of creating a technical drawing with the use of computer software. CAD software is used to increase the productivity of the designer, improve the quality of design, improve communications through documentation, and to create a database for manufacturing. CAD output is often in the form of electronic files for print or machining operations. CAD software uses either vector based graphics to depict the objects of traditional drafting, or may also produce raster graphics showing the overall appearance of designed objects.

CAD often involves more than just shapes. As in the manualdrafting of technical and engineering drawings, the output of CAD must convey information, such as materials, processes, dimensions, and tolerances, according to application-specific conventions. CAD may be used to design curves and figures in two-dimensional (2D) space; or curves, surfaces, and solids in three-dimensional (3D) space.

CAD is an important industrial art extensively used in many applications, including automotive, shipbuilding, and aerospace industries, industrial and architectural design, prosthetics, and many more. CAD is also widely used to produce computer animation for special effects in movies, advertising and technical manuals. The modern ubiquity and power of computers means that even perfume bottles and shampoo dispensers are designed using techniques unheard of by engineers of the 1960s. Because of its enormous economic importance, CAD has been a major driving force for research in computational geometry, computer graphics (both hardware and software), and discrete differential geometry.

In CAD, many commands are available for drawing basic geometric shapes. Examples include CIRCLE, POLYGON, ARC, ELLIPSE, and more.

4. Reliability design

Reliability engineering is an engineering field that deals with the study, evaluation, and life-cycle management of reliability: the ability of a system or component to perform its required functions under stated conditions for a specified period of time. Reliability engineering is a sub-discipline within systems engineering. Reliability is often measured as probability of failure, frequency of failures, or in terms of availability, a probability derived from reliability and maintainability. Maintainability and maintenance are often important parts of reliability engineering.

Reliability engineering is closely related to safety engineering, in that they use common methods for their analysis and may require input from each other. Reliability engineering focuses on costs of failure caused by system downtime, cost of spares, repair equipment, personnel and cost of warranty claims. The focus of safety engineering is normally not on cost, but on preserving life and nature, and therefore deals only with particular dangerous system failure modes.

5. Sustainable design

Sustainable design (also called environmental design, green design, environmentally sustainable design, ecological design etc.) is the philosophy of designing physical objects, the built environment, and services to comply with the principles of social, economic and ecological, sustainability.

The intention of sustainable design is to "eliminate negative environmental impact completely through skillful, sensitive design". Manifestations of sustainable design require no non-renewable resources, impact the environment minimally, and relate people with the natural environment.

Beyond the "elimination of negative environmental impact", sustainable design must create projects that are meaningful innovations that can shift behaviour. A dynamic balance between economy and society, intended to generate long-term relationships between user and object/service and finally to be respectful and mindful of the environmental and social differences.

6. Design methodology

The goal of design methods is to gain key insights or unique essential truths resulting in more holistic solutions in order to achieve better experiences for users with products, services, environments and systems they rely upon. Insight, in this case, is clear and deep investigation of a situation through design methods, thereby grasping the inner nature of things intuitively.

Design methodology is a comprehensive subject research on product design rules, design procedure, design thinking and work method. The design methodology analyses design strategy and processes, as well as tactical issues of design means and methods in view of system engineering, promotes the comprehensive use of modern design theory, scientific methods, advanced means and tools in research. It has a positive effect on the development of new products, the transformation of the old products and the improvement of the products' competitiveness in the market.

New Words and Expressions

1. emerging 新兴的
2. optimal design 优化设计
3. computer aided design 计算机辅助设计
4. reliability calculation 可靠性计算
5. optimization theory 最优化理论
6. design variable 设计变量
7. objective function 目标函数
8. constrained condition 约束条件
9. optimal scheme 最优方案
10. hydrodynamic journal bearing 滑动轴承
11. non-linear programming 非线性规划
12. mathematical model 数学模型
13. tribology design 摩擦学设计
14. embody 体现
15. astronautic 航天学
16. seismology 地震学
17. interdisciplinary 跨学科的
18. dry contact 干接触
19. boundary contact 边界接触
20. mixed contact 混合接触
21. thin film lubrication 薄膜润滑
22. elastohydrodynamic lubrication(EHL) 弹性流体动力润滑
23. hydrodynamic lubrication 流体动力润滑
24. hydrostatic lubrication 流体静力润滑
25. productivity 生产率
26. documentation 文献资料
27. database 数据库
28. electronic file 电子文件
29. automotive 汽车的
30. shipbuilding 造船业
31. prosthetics 修复
32. ubiquity 普遍存在
33. shampoo dispenser 洗发水喷头
34. reliability engineering 可靠性工程
35. probability 可能性
36. availability 有效性
37. maintainability 保持性
38. sustainable design 可持续设计
39. philosophy 哲学
40. non-renewable resources 不可再生资源
41. design methodology 设计方法学
42. holistic 全面的
43. collaboration 合作
44. by nature 本质上

❖ Writing Training

英汉互译技巧：词量的增减

一、增译法

根据英汉两种语言不同的思维方式、语言习惯和表达方式，在翻译时常增添一些词、短句或句子，以便更准确地表达出原文所包含的意义，这种方式多半用在汉译英里。

汉语无主句较多，而英语句子一般都要有主语，所以在翻译汉语无主句的时候，除了少数可用英语无主句外，一般用被动语态或补出主语，使句子完整。

英汉两种语言在名词、代词、连词、介词和冠词的使用方法上也存在很大差别。英语中代词使用频率较高，凡说到人的器官和归某人所有的或与某人有关的事物时，必须在前面加上物主代词。因此，在汉译英时需要增补物主代词，而在英译汉时又需要根据情况适当地删减。

英语词与词、词组与词组以及句子与句子的逻辑关系一般用连词来表示，而汉语则往往通过上下文和语序来表示这种关系。因此，在汉译英时常常需要增补连词。英语句子离不开介词和冠词。

在汉译英时还要注意增补一些原文中暗含而没有明言的词语和一些概括性、注释性的词语，以确保译文意思的完整。

［例1］ Indeed, the reverse is true.

译文：实际情况恰好相反。（增译名词）

［例2］ 这是这两代计算机之间的又一个共同点。

译文：This is yet another common point between the computers of the two generations. （增译介词）

［例3］ Individual mathematicians often have their own way of pronouncing mathematical expressions and in many cases there is no generally accepted "correct" pronunciation.

译文：各个数学家对数学公式常常有自己的读法，在许多情况下，并不存在一个普遍接受的所谓"正确"读法。（增加隐含意义的词）

［例4］ 只有在可能发生混淆或要强调其观点时，数学家才使用较长的读法。

译文：It is only when confusion may occur, or where *he/she* wishes to emphasis the point, that the mathematician will use the longer forms. （增加主语）

二、省译法

这是与增译法相对应的一种翻译方法，即删去不符合目标语思维习惯、语言习惯和表达方式的词，以避免译文累赘。

［例1］ You will be staying in this hotel during *you* visit in Beijing.

译文：你在北京访问期间就住在这家饭店里。（省译物主代词）

［例2］ I hope you will enjoy your stay here.

译文：希望您在这儿过得愉快。（省译主语）

［例3］ 中国政府历来重视环境保护工作。

译文：The Chinese government has always attached great importance to environmental protection. （省译名词）

［例4］ The development of IC made it possible for electronic devices to become smaller and smaller.

译文：集成电路的发展使得电子器件可以做得越来越小。（省译形式主语 it）

Reading Material

Mechanical Design

Design is the first step in the manufacture of any product. A new machine is born because there is a real or imagined need for it. It evolves from designer's conception of a device with which to accomplish a special purpose. Mechanical design means the design of things and systems of a mechanical nature-machines, products, structures, devices, and instruments. For the most part mechanical design utilizes mathematics, the materials sciences, and the engineering-mechanics sciences.

The total design process is of interest to us. How does it begin? Does the engineer simply sit down at his desk with a blank sheet of paper? And, as he jots down some ideas, what happens next? What factors influence or control the decisions which have to be made? Finally, then, how does this design process end?

Sometimes, but not always, design begins when an engineer recognizes a need and decide to do something about it. Recognition of the need and phrasing it in so many words often constitute a highly creative act because the need may be only a vague discontent, a feeling of uneasiness, or a sensing that something is not right.

The need is usually not evident at all. For example, the need to do something about a food-packaging machine may be indicated by the noise level, by the variation in package weight, and by slight but perceptible variations in the quality of the packaging or wrap.

There is a distinct difference between the statement of the need and the identification of the problem which follows this statement. The problem is more specific. If the need is for cleaner air, the problem might be that of reducing the dust discharge from power-plant stacks, or reducing the quantity of irritants from automotive exhausts.

Definition of the problem must include all the specifications for the thing that is to be designed. The specifications are the input and output quantities, the characteristics and dimensions of the space the thing must occupy and all the limitations on these quantities. We can regard the thing to be designed as something in a black box. In this case we must specify the inputs and outputs of the box together with their characteristics and limitations. The specifications define the cost, the number to be manufactured, the expected life, the range, the operating temperature, and the reliability.

There are many implied specifications which result either from the designer's particular environment or from the nature of the problem itself. The manufacturing processes which are available, together with the facilities of a certain plant, constitute restrictions on a designer's freedom, and hence are a part of the implied specifications. A small plant, for instance, may not own cold-working machinery. Knowing this, the designer selects other metal-processing methods which can be performed in the plant. The labor skills available and the competitive situation also constitute implied specifications.

After the problem has been defined and a set of written and implied specifications has been obtained, the next step in design is the synthesis of an optimum solution. Now synthesis can not take place without both analysis and optimization because the system under design must be analyzed to determine whether the performance complies with the specifications.

The design is an iterative process in which we proceed through several steps, evaluate the results, and then return to an earlier phase of the procedure. Thus we may synthesize several components of a system, analyze and optimize them, and return to synthesis to see what effect this on the remaining parts of the system. Both analysis and optimization require that we construct or devise abstract models of the system which will admit some form of mathematical models. In creating them it is our hope that we can find one that will simulate the real physical system as well.

New Words and Expressions

1. recognition 识别
2. uneasiness 不自在
3. identification 确认
4. black box 黑箱
5. metal-processing 金属加工
6. synthesis 综合,合成
7. iterative process 迭代过程

Lesson 15　Computer-Integrated Manufacturing

Text

Manufacturing industries strive to reduce the cost of the product continuously to remain competitive in the face of global competition. In addition, there is the need to improve the quality and performance levels on a continuing basis. Another important requirement is on time delivery. In the context of global outsourcing and long supply chains cutting across several international borders, the task of continuously reducing delivery times is really an arduous task.

In order to meet the above requirements, manufacturing industries use automatic devices and controls in mechanized production lines increasingly. The advances in automation have enabled industries to develop islands of automation such as computer-aided design(CAD), computer-aided manufacturing (CAM), flexible manufacturing system (FMS), automatic storage and retrieval system (AS/RS), robotics, total quality management, manufacturing resource planning (MRP), office automation, and so on. However, communication among these islands was once handled manually. This limited the level of improvement in productivity that could be accomplished in the overall manufacturing process. Computer-integrated manufacturing (CIM) may be viewed as the successor technology which links these islands and other automated components of manufacturing to overcome these limitations.

CIM makes full use of the capabilities of the digital computer to improve manufacturing. Two of them are: (1) Variable and programmable automation. (2) Real time optimization. By using powerful computer systems to integrate all phases of manufacturing(see Figure 15.1), from initial customer order to final shipment, firms hope to increase productivity, improve quality, meet customer needs faster, and offer more flexibility. This requires sharing of information among different applications or sections of a factory, accessing incompatible and heterogeneous data and devices. For example, the product data is created during design. This data has to be transferred from the modeling software to manufacturing software without any loss of data. CIM uses a common database wherever feasible and communication technologies to integrate design, manufacturing and associated business functions. The data required for various functions are passed from one application software to another in a seamless manner. The capabilities of the computer are thus exploited not only for the various bits and pieces of manufacturing activity but also for the entire system of manufacturing. CIM reduces the human component of manufacturing and thereby relieves the process of its slow, expensive and error-prone component.

Lesson 15 Computer-Integrated Manufacturing

Figure 15.1 Main elements of a CIM system

Less comprehensive computerized systems for production planning, inventory control, or scheduling are often considered part of CIM. Some factors involved when considering a CIM implementation are the production volume, the experience of the company or personnel to make the integration, the level of the integration into the product itself and the integration of the production processes. Nowadays, less than 1 percent of U. S. manufacturing companies have approached full-scale use of CIM, but more than 40 percent are using one or more elements of CIM technology. Some studies asked managers how much their companies invest in several of the technologies that comprise CIM by measuring investment on a 7-point scale (1 = no investment and 7 = heavy investment). Computer-aided design received the highest average score (5. 2), followed by numerically controlled machines (4. 8), computer-aided manufacturing (4. 0), flexible manufacturing systems (2. 5), automated materials handling (2. 3), and robots (2. 1).

There are three major challenges to development of a smoothly operating computer-integrated manufacturing system:

• Integration of components from different suppliers: When different machines, such as CNC, conveyors and robots, are using different communications protocols. In the case of AGVs, even lengths of time for charging the batteries may cause problems.

• Data integrity: The higher the degree of automation, the more critical is the integrity of the data used to control the machines. A large amount of operational data and information from different sensors have to be collected and processed in time with useful knowledge and information acquired. While the CIM system saves on labor of operating the machines, it requires extra human labor in ensuring that there are proper safeguards for the data signals that are used to control the machines.

• Process control: Computers may be used to assist the human operators of the manufacturing facility, but there must always be a competent engineer on hand to handle circumstances which could not be foreseen by the designers of the control software. It is becoming increasingly difficult for operators to understand the many signals and information from display board, video, and computer screen.

New Words and Expressions

1. Computer-integrated manufacturing(CIM) 计算机集成制造
2. outsourcing   外部采购
3. arduous   费劲的,险峻的
4. island   孤岛,孤立的事物
5. automatic storage and retrieval systems (AS/RS)
   自动存储和检索系统
6. manufacturing resource planning (MRP)
   制造资源计划
7. successor   后继,继任者
8. variable   变量
9. incompatible   不相容的,矛盾的
10. heterogeneous   异类的,不同的
11. database   数据库
12. inventory   清单,商品目录
13. protocol   草案,协议
14. AGV (the automatic guided vehicle)
    自动导引车系统
15. integrity   完整性,完全

Notes

(1) Some factors involved when considering a CIM implementation are the production volume, the experience of the company or personnel to make the integration, the level of the integration into the product itself and the integration of the production processes.

在考察计算机集成制造系统的实施情况时,会考虑这些因素:产量、企业或个人的集成经验、产品本身的集成水平和生产过程的集成水平。

(2) Some studies asked managers how much their companies invest in several of the technologies that comprise CIM by measuring investment on a 7-point scale (1 = no investment and 7 = heavy investment).

一些研究就企业在构成计算机集成制造系统的几项技术上的投资情况对其经营者进行了调查,并通过7分制来衡量(1分＝没有投资,7分＝大力投资)。

❖ Writing Training

操作过程描述

在英文科技写作中,有时需要对操作过程进行描述,来引导读者按照一定的思维逻辑循序渐进,知道发生了什么、怎么发生的,或者该做什么、怎么做。这些内容带有描述说明的性质,客观而不带有感情色彩,具有准确、简明、客观的特点,这就要求在英译过程中注意以下几点。

1. 为了把信息内容如实准确地翻译出来,一些专业术语、固定用语和习惯说法必须表达得准确、地道,且常用缩略形式,例如发光二极管(light emitting diode)缩写成LED,计算机辅助制造(computer-aided manufacturing)缩写成CAM,计算机数字控制机床(computer numerical control)缩写成CNC。

2. 常使用非人称名词化结构作主语,使句意更客观、简洁。

[例] 人们应用电子计算机,促使劳动生产率突飞猛进。

译文:The application of electronic computers makes for a tremendous rise in labor productivity.

3. 普遍使用一般现在时。一般现在时可以用来表示不受时间限制的客观存在,包括客观真理、格言、科学事实及其他不受时限的事实。操作过程往往是无时间性的一般叙述,其译文普遍使用一般现在时,以体现出内容的客观性和形式的简明性。

4. 常使用被动语态。操作过程强调的是所叙述的事物本身,而并不需要过多地注意它的行为主体(即施动者)。基于这样的特点,在英译过程中大量使用被动语态,以使译文客观简洁。例如:

[例] 操作者将试样的一端固定,在另一端加载。

译文:The test specimen is fixed at one end and loaded at the other.

5. 广泛使用祈使句。操作过程很多时候是指导使用者要做什么,不要做什么或该怎么做,不需要过多地注意它的行为主体,所以其译文也经常使用祈使句,谓语一般用动词原形,省略主语,译文的表述显得客观、简洁。

以下是某显示器安装过程的英文描述。

(1) Turn off your computer and unplug its power cable.

(2) Connect the blue connector of the video cable to the blue video connector on the back.

(3) Connect your monitor's power cable to the power port on the back of the monitor.

(4) Plug your computer's power cord and your monitor into a nearby outlet.

(5) Turn on your computer and monitor. If the monitor displays an image, installation is complete.

6. 内容条目简洁明了,步骤清晰,逻辑性强。

通常操作过程的描述可以按先后顺序来写,表达先后顺序的方式如下:

(1) 简单地以数字序号(1、2、3……或者 STEP1、STEP2、STEP3……)来表示,如上例所示。

(2) 以操作流程图来表示。

(3) 用表示次序或时间关系的连接词,使表达更顺畅。常见的连接词有:

First(ly), second(ly), furthermore(此外,而且), finally;

First and most important(首先也是最重要), last but not least(最后但并非最不重要的);

To begin/start with(首先,以……开始), next, then, afterwards, lastly;

Before, after, moreover(再者), eventually.

这些词可以相互搭配,灵活运用。除了使用连接词外,配以空间的变化也可以使描述更清晰。以下的例子描述了产品从零件生产到装配成最终产品的过程:

In the part fabrication center, raw material is transformed into piece parts. *Afterwards*, some piece parts move by robot carrier or automatic guided vehicle to the component fabrication center. Here materials handlers of various types, and other reprogrammable automation, put piece parts together to form components. Components may *then* be transferred to the product assembly center for final assembly operations there. The *final* product moves out of the product assembly center to the product distribution center or in some cases directly to the end user.

以下的例子说明了模型飞机的安装制作过程：

At first you must cut the parts off from the plastic mould plates. Be careful not to break the delicate parts. Do not use too much glue, or it will spoil the surface when you glue together the parts according to the instructions. *Then* you can paint them. Painting is the most important step. It makes the plane more like a real one. *At last* a delicate plane will be born in your hands.

读者可以练习用不同的连接词（也可改变连接词位置）和句式（如被动句或祈使句）来重新改写这一制作过程。

❖ Reading Material

Computer-Integrated Manufacturing

Since about 1970 there has been a growing trend in manufacturing firms toward the use of computers to perform many of the functions related to design and production. The technology associated with this trend is called CAD/CAM. Computer-aided design (CAD) makes use of computer systems to assist in the creation, modification, analysis, and optimization of a design. Computer-aided manufacturing (CAM) deals directly with manufacturing operations. CAM systems are used to design production processes and to control machine tools and materials flow through programmable automation. A CAD/CAM system integrates the design and manufacturing function by translating final design specifications into detailed machine instructions for manufacturing an item. It is quicker, less error prone than humans, and eliminates duplication between engineering and manufacturing.

Today it is widely recognized that the scope of computer applications must extend beyond design and production to include the business functions of the firm. These business functions include order entry, cost accounting, employee time records and payroll, and customer billing. The name given to this more comprehensive use of computers is computer-integrated manufacturing (CIM). CIM is considered a natural evolution of the technology of CAD/CAM which by itself evolved by the integration of CAD and CAM. In an ideal CIM system, computer technology is applied to all the operational and information-processing functions of the company, from customer orders through design and production (CAD/CAM) to product shipment and customer service. In many ways, CIM represents the highest level of automation in manufacturing.

There is an instance of computer integrated manufacturing in a factory which manufactures DVD/CD storage units. All functions are under computer control. This starts with computer aided design, followed by computer aided manufacture, then followed by automated storage and distribution. One integrated computer system controls all that happens (Figure 15.2).

Stage one—Computer aided design. A product is designed and tested totally on computer

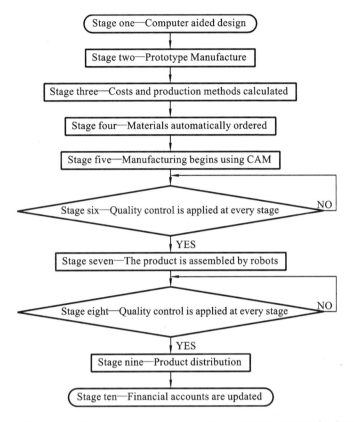

Figure 15.2 An integrated computer system controls all functions of a factory

by using CAD software.

Stage two—Prototype manufacture. Prototypes are manufactured on machines such as 3D printers which produce an accurate 3D model. CNC routers and laser cutters may also be used to produce a realistic model. Sometimes working models are manufactured.

Stage three—The computer system controlling the plant works out the most efficient method of manufacture. It calculates costs, production methods, numbers to be manufactured, storage and distribution.

Stage four—The computer system orders the necessary materials to manufacture the product, keeping costs to a minimum. The "just in time" philosophy is applied. This means that materials/components are ordered as needed. Very little is stored at the factory. Usually only enough materials are stored to keep the factory going for a small number of days. Materials are automatically reordered when required, to keep the factory working smoothly and continuously.

Stage five—Manufacturing begins with the product being made using CAM. Computers control CNC machines such as laser cutters, CNC routers and CNC lathes.

Stage six—Quality control is applied at every stage. The product is tested using computer control inspections. For instance, the accuracy of manufacture is tested automatically. This ensures that the product is manufactured to the correct sizes.

Stage seven—The product is assembled by robots. This is automated (controlled) by the computer system.

Stage eight—The product is quality checked before being stored for distribution to the customer. All storage is automated. This means that computer controlled vehicles move the finished product from the manufacturing area to storage. The computer systems keep track of every individual product. Products are bar coded which are constantly scanned and recorded by the computer system.

Stage nine—The product is automatically moved from store to awaiting lorries/trucks for distribution to the customer.

Stage ten—Financial accounts are updated, bills chased up and paid by the computer system.

New Words and Expressions

1. computer-aided design(CAD) 计算机辅助设计
2. storage unit 存储单元
3. prototype 原型
4. computer-aided manufacturing(CAM) 计算机辅助制造
5. 3D printer 三维打印机
6. duplication 重复,副本
7. materials flow 物料流
8. router 刨槽机,缩放刻模机,刨圆削片联合机
9. order entry 订单登记(录)
10. laser cutter 激光切割机
11. cost accounting 成本核算
12. bar code 条形码
13. customer billing 给顾客开出账单
14. distribution 分配,配送

Lesson 16　Industrial Robots

Text

Robots are designed for many purposes. They are deployed for volcanic examination, underwater mineral deposits discovery, fire and bomb fighting, nuclear site inspection, home cleaning, laboratory research, medical surgery, office mail delivery as well as a myriad of other tasks. Today, most robots are used in industry manufacturing. Robots can be similar in form to a human, but industrial robots do not resemble people at all. The most widely accepted definition of an industrial robot is one developed by the Robotic Industries Association: An industrial robot is a reprogrammable, multifunctional manipulator designed to move materials, parts, tools, or specialized devices through variable programmed motions for the performance of a variety of tasks. In the context of general robotics, most types of industrial robots would fall into the category of robotic arms(see Figure 16.1).

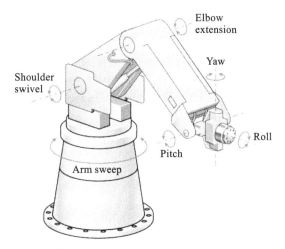

Figure 16.1　Industrial robot and its standard movements

The manipulator can be divided into two sections: (1) an arm-and-body, which usually consists of three joints connected by large links, and (2) a wrist, consisting of two or three compact joints. The two manipulator sections have different functions: the arm-and-body is used to move and position parts or tools in the robot's work space, while the wrist is used to orient the parts or tools at the work location. Attached to the wrist is the end-effecter to perform task, and therefore is also known as a robot tool. It can be a welding head, a spray gun, a machining tool, or a gripper containing on-off jaws, depending upon the specific application of the robot.

1. Characteristics of industrial robots

• Number of axes—two axes are required to reach any point in a plane; three axes are required to reach any point in space. To fully control the orientation of the end of the arm (i.e. the wrist) three more axes (yaw, pitch, and roll) are required.

• Working envelope—the region of space a robot can reach.

• Kinematics—the actual arrangement of rigid members and joints in the robot, which determines the robot's possible motions.

• Carrying capacity or payload—how much weight a robot can lift.

• Speed—how fast the robot can position the end of its arm.

• Accuracy—how closely a robot can reach a commanded position. When the absolute position of the robot is measured and compared to the commanded position the error is a measure of accuracy. Accuracy can be improved with external sensing for example a vision system or Infra-Red.

• Repeatability—how well the robot will return to a programmed position. This is not the same as accuracy. It may be that when told to go to a certain x-y-z position that it gets only to within 1 mm of that position. This would be its accuracy which may be improved by calibration. But if that position is taught and teaching information is stored in controller memory and each time the robot is sent there it returns to within 0.1 mm of the taught position then the repeatability will be within 0.1 mm.

Accuracy and repeatability are different measures. Repeatability is usually the most important criterion for a robot.

• Power source—some robots use electric motors, others use hydraulic actuators. The former are faster, the latter are stronger and advantageous in applications such as spray painting, where a spark could set off an explosion.

• Drive—some robots connect electric motors to the joints via gears; others connect the motor to the joint directly (*direct drive*).

2. Robot programming

The computer system that controls the manipulator must be programmed to teach the robot the particular motion sequence and other actions that must be performed in order to accomplish its task. There are several ways that industrial robots are programmed.

One method is called teaching programming. This requires that the manipulator be driven through the various motions needed to perform a given task, recording the motions into the robot's computer memory. This can be done either by physically moving the manipulator through the motion sequence or by using a control box (e.g. teach pendant, a handheld control and programming unit capable to manually send the robot to a desired position) to drive the manipulator through the sequence.

A second method of programming involves the use of a programming language very much like a computer programming language. Specialized robot software is run either in the robot controller or in the computer linked with controller or both depending on the system design. In addition to many of the capabilities of a computer programming language (i.e., data processing, computations, communicating with other computer devices, and decision making), the robot language also includes statements specifically designed for robot control. These capabilities include motion control and input/output. Motion-control commands are used to direct the robot to move its manipulator to some defined position in space. For example, the statement "move P1" might be used to direct the robot to a point in space called P1. Input/output commands are employed to control the receipt of signals from sensors and other devices in the work cell and to initiate control signals to other pieces of equipment in the cell. For instance, the statement "signal 3, on" might be used to turn on a motor in the cell, where the motor is connected to output line 3 in the robot's controller.

The teach pendant or PC is usually disconnected after programming and the robot then runs on the program that has been installed in its controller.

New Words and Expressions

1. deploy   使用,调度,配置
2. mineral deposit   矿床
3. a myriad of   无数的,多种
4. resemble   像,类似
5. Robotic Industries Association   机器人工业协会
6. manipulator   机械手
7. joint   接头,关节,接合处
8. link   连杆,连接物
9. compact   紧凑的,紧密的
10. end-effector   末端执行器
11. gripper   钳子,手爪
12. yaw   偏转
13. pitch   倾斜
14. roll   转动
15. working envelope   工作范围
16. kinematics   运动学
17. articulated   有关节的,铰接的
18. Infra-Red   红外线(的)
19. hydraulic actuator   液压致动器
20. teach pendant   示教器
21. work cell   工作间
22. initiate   开始,发动,发起

Notes

(1) The most widely accepted definition of an industrial robot is one developed by the Robotic Industries Association: An industrial robot is a reprogrammable, multifunctional manipulator designed to move materials, parts, tools, or specialized devices through variable programmed motions for the performance of a variety of tasks. In the context of general robotics, most types of industrial robots would fall into the category of robotic arms.

最为广泛接受的工业机器人的定义是由机器人工业协会提出的:工业机器人是可重复编程、多功能的机械手,其设计目的是通过可改变的编程动作去移动材料、零件、工具或专用设备,来执行不同的任务。就常规的机器人技术而言,大部分类型的工业机器人都属于机械手的范畴。

(2) It may be that when told to go to a certain x-y-z position that it gets only to within

1 mm of that position. This would be its accuracy which may be improved by calibration. But if that position is taught and teaching information is stored in controller memory and each time the robot is sent there it returns to within 0.1 mm of the taught position then the repeatability will be within 0.1 mm.

当被指示要到达一个由 x,y,z 坐标确定的点时,它(机器人)到达了距离该点 1 mm 以内的位置,这是精确度,精确度可以通过校正来改进。但是如果这个位置被示教并且示教信息被储存于控制器内存中,每次机器人被送往那里,它都会回到距离示教位置 0.1 mm 以内的地方,则重复性就为 0.1 mm。

❖ Writing Training

如何写英文科技论文(一)

英文科技论文写作是进行国际学术交流必需的技能。在撰写英语科技论文时,除了遵循科技论文的基本要求如"讲究逻辑,表达清晰,用词准确"外,还需要注意英文科技论文的写作格式。究其原因,主要在于中国人和西方人思维方式有所不同,从而导致了写作风格的差异。这一点突出表现在文章结构上和表达上的不同。譬如,通常中国人行文较为含蓄,因此文章各段之间可能存在不明显的内在关联,而西方人则比较直截了当,他们的文章结构往往一目了然。因此,即使已有一篇现成的中文论文,在其基础上写英文论文时也不能采用"拿来主义",逐字逐句翻译。

撰写英文科技论文的第一步就是推敲结构,使之具备西方人易于理解的形式。最简单有效的方法即采用 IMRaD 形式(Introduction, Materials and Methods, Results, and Discussion)——西方科技论文最通用的一种结构方式。

IMRaD 结构的逻辑性体现在它能依次回答以下问题。

- 研究的是什么问题?——答案就是 Introduction。
- 这个问题是怎么研究的?——答案就是 Materials and Methods。
- 发现了什么?——答案就是 Results。
- 这些发现意味着什么?——答案就是 Discussion。

如果按照这个结构整体规划论文,有一个方法值得借鉴,即剑桥大学爱希比教授提出的"概念图"(见图 16.2)。首先在一张大纸上(A4 或 A3 纸,横放)写下文章题目(事先定好题目很重要),然后根据 IMRaD 的结构确定基本的段落主题,把它们写在不同的方框内。你可以记录任何你脑海中闪现的可以包括在该部分之内的内容,诸如段落标题、图表、需要进一步阐述的观点等等,把它们写在方框附近的圈内,并用箭头标示它们所属的方框。画概念图的阶段也是自由思考的阶段,在此过程中不必拘泥于细节。哪些东西需要包括进文章?还需要做哪些工作——找到某文献的原文,补画一张图表,还是再查找某个参考文献?当你发现自己需要再加进一个段落时,那就在概念图中添一个新的方框。如果你发现原来的顺序需作调整,那就用箭头标示新的顺序。绘制概念图的过程看似儿童游戏,但其意义重大,它可以给你自由思考的空间,并通过图示的方式记录你思维发展的过程。这便是写论文的第一步:从整体考虑文章结构,思考各种组织文章的方法,准备好所需的资料,随时记录出现的新想法。采用这个方法,不论正式下笔时是从哪一部分写起,都能够做到有条不紊。以下对论文的基本组成部分及需要特别注意的方面逐一进行阐述。

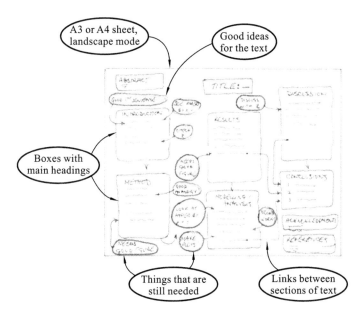

图 16.2 爱希比教授用来规划文章的"概念图"

❖ Reading Material

Application of Industrial Robots

The first truly modern robot, digitally operated and programmable, was invented by George Devol in 1954 and was ultimately called the Unimate. Devol sold the first Unimate to General Motors in 1960, and it was installed in 1961 in a plant in Trenton, New Jersey to lift hot pieces of metal from a die casting machine and stack them. Devol's patent for the first digitally operated programmable robotic arm represents the foundation of the modern robotics industry. Since then, the industrial robot has blossomed, mainly in the US, Japan and Europe. Japanese companies which enjoyed the deeper long term financial resources and strong domestic market because of a labor shortage eventually prevailed. Their robots spread all over the globe. Only a few non-Japanese companies managed to survive in this market.

There are certain characteristics of industrial jobs performed by humans that identify the work as a potential application for robots: (1) the operation is repetitive, involving the same basic work motions every cycle; (2) the operation is hazardous or uncomfortable for the human worker ; (3) the task requires a work part or tool that is heavy and awkward to handle; and (4) the operation allows the robot to be used on two or three shifts. The applications can be divided into three categories: (1) material handling, (2) processing operations, and (3) assembly and inspection.

Material-handling applications include material transfer and machine loading and unloading. Material-transfer applications require the robot to move materials or work parts from one location to another. Many of these tasks are relatively simple, requiring robots to pick up parts from one conveyor and place them on another. Other transfer operations are more complex, such as placing parts onto pallets in an arrangement that must be calculated by the robot. Machine loading and unloading operations utilize a robot to load and unload parts at a production machine. This requires the robot to be equipped with a gripper that can grasp parts. Usually the gripper must be designed specifically for the particular part geometry.

In robotic processing operations, the robot manipulates a tool to perform a process on the work part. Examples of such applications include spot welding, continuous arc welding, laser welding and spray painting. Spot welding of automobile bodies is one of the most common applications of industrial robots in the United States. The robot positions a spot welder against the automobile panels and frames to complete the assembly of the basic car body. Arc welding is a continuous process in which the robot moves the welding rod along the seam to be welded. Laser welding is starting to become more widespread for welding, as a means of replacing spot welding. The main advantage of laser welding is that it requires access from only one side. Spray painting involves the manipulation of a spray-painting gun over the surface of the object to be coated. Other operations in this category include grinding, polishing, and routing, in which a rotating spindle serves as the robot's tool.

The third application area of industrial robots is assembly and inspection. The use of robots in assembly is expected to increase because of the high cost of manual labour common in these operations. Since robots are programmable, one strategy in assembly work is to produce multiple product styles in batches, reprogramming the robots between batches. An alternative strategy is to produce a mixture of different product styles in the same assembly cell, requiring each robot in the cell to identify the product style as it arrives and then execute the appropriate task for that unit. Inspection is another area of factory operations in which the utilization of robots is growing. In a typical inspection job, the robot positions a sensor with respect to the work part and determines whether the part is consistent with the quality specifications.

Robots are the drudges of the workforce, performing highly repetitive tasks without tiring, taking a lunch break, or complaining. Benefits from robot installation include less waste materials, more consistent quality, increased productivity and shorter workweeks for labour.

However, unsafe use of robots constitutes an actual danger. A heavy industrial robot with powerful actuators and unpredictably complex behavior can cause harm, for instance by stepping on a human's foot or falling on a human. Most industrial robots operate inside a security fence which separates them from human workers, but not all. There were four robot-caused deaths.

Of particular concern for many labour specialists is the impact of industrial robots on the work force, since robot installations involve a direct substitution of machines for humans, sometimes at a ratio of ten humans per robot. The long-term effects of robotics on employment and unemployment rates are debatable. Most studies in this area have been controversial and inconclusive. Some analysts argue that robots and other forms of automation will ultimately result in significant unemployment as machines begin to match and exceed the capability of workers to perform most jobs. At present the negative impact is only on menial and repetitive jobs, and there is actually a positive impact on the number of jobs for highly skilled technicians, engineers, and specialists. However, these highly skilled jobs are not sufficient in number to offset the greater decrease in employment among the general population, causing structural unemployment in which overall (net) unemployment rises.

These dangers aside, robotics, if used wisely and effectively, can relieve humans from repetitive, hazardous, and unpleasant labour in all forms and provide a growing social and economic environment in which humans can enjoy a higher standard of living and a better way of life in the future.

New Words and Expressions

1. Unimate 通用机械手（一种机器人的商品名）
2. shift 轮班
3. General Motor 通用汽车公司
4. conveyor 传送带,输送机
5. die casting 压力铸造
6. pallet 货盘,平板架
7. geometry 几何学
8. grinding 研磨,磨削
9. spot welding 点焊
10. polishing 抛光
11. continuous arc welding 连续电弧焊
12. routing 走线
13. laser welding 激光焊接
14. batch 批,成批
15. inspection 检查,观测
16. unemployment rate 失业率
17. sensor 传感器
18. ratio 比,比率
19. drudge 苦工,做单调无味或卑贱工作的人
20. controversial 争论的,争议的
21. actuator 执行器,拖动装置,操作机构
22. inconclusive 非决定性的,不确定的
23. spray painting 喷漆

Lesson 17 Wastewater Treatment Facilities

Text

The term wastewater is commonly used to describe liquid wastes that are collected and transported to a treatment facility through a system of sewers. Wastewater is generally divided into two broad classifications: domestic wastewater and industrial wastewater. Wastewater treatment is the process of removing suspended solids, undesirable chemicals, biological contaminants and gases from contaminated water. The processes generally consist of primary, secondary, or tertiary treatment. A primary clarifier system is used to remove solid and floating materials and to prepare wastewater for biological treatment. Secondary treatment systems utilize aerobic microorganisms in biological reactors to remove dissolved organic materials. The tertiary system, or advanced treatment system, is becoming more prevalent to remove specific residual substances, trace organic materials, and to disinfect the water before discharge to a surface water stream or ocean outfall. In addition, evaporation, distillation, electrodialysis, ultrafiltration, reverse osmosis, freeze drying, freeze thaw, floatation, and land application, with particular emphasis on the increased use of natural and constructed wetlands, are being studied and utilized as methods for advanced wastewater treatment. Typical facilities and equipment used for wastewater treatment are as follows:

Sedimentation Tanks Particles that experience a force, either due to gravity or due to centrifugal motion will tend to move in a uniform manner in the direction exerted by that force. For gravity settling, the particles will tend to fall to the bottom of the vessel, forming slurry at the vessel base. In the primary treatment stage, wastewater flows through sedimentation tanks that are used to settle sludge while grease and oils rise to the surface and are skimmed off (see Figure 17.1). Primary settling tanks are usually equipped with mechanically driven scrapers which can continually drive the collected sludge towards a

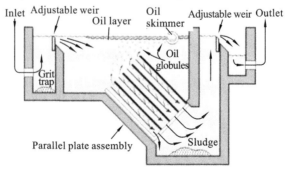

Figure 17.1 Schematic diagram of a sedimentation tank

hopper in the base of the tank where it is pumped to sludge treatment facilities.

Biological reactors: Many different types of reactors are used in environmental engineering practices. Individual reactors are generally designed to emphasize suspended-growth or fixed-films. Suspended-growth systems include activated sludge, where the biomass is mixed with biodegradable organic compounds in the wastewater. Fixed-film or attached growth systems include trickling filters, bio-towers, and rotating biological contactors, etc, where the biomass grows on media and the wastewater passes over its surface. Factors influencing the choice among the different reactor types can include: the physical and chemical characteristics of the waste being considered, the concentration of contaminants being treated, the presence or absence of oxygen, the efficiency of treatment and system reliability required, the climatic conditions under which the reactor will operator, the number of different biological processes involved in the overall treatment system, the skills and experience of those who will operate the system, and the relative costs at a given location and time for construction and operation of different possible reactor configurations. Typical reactors used in environmental applications are illustrated in Figure 17.2.

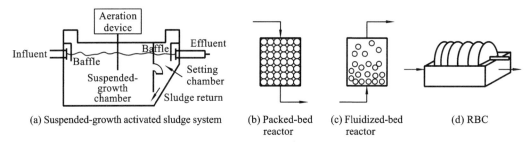

(a) Suspended-growth activated sludge system   (b) Packed-bed reactor   (c) Fluidized-bed reactor   (d) RBC

Figure 17.2 Typical biological reactors

Packed bed reactors   The most common fixed-film reactor is the packed bed. Historically, large rocks have been used as support media, but today it is more common to use plastic media or pea-sized stones. Both are light and offer greater surface area and pore volume per unit of reactor volume than do large rocks. Commonly, packed-bed reactors are known as trickling filters or biological towers. Here, the wastewater is distributed uniformly over the surface of the bed and allowed to trickle over the surface of the media, giving the packed-bed reactors some plug-flow character. The void space remains open to the passage of air so that oxygen can be transferred to the microorganism through the reactor.

Fluidized-bed reactors   The fluidized-bed reactor depends upon the attachment of microorganisms to partials that are maintained in suspend by a high upward flow rate of the fluid to be treated. The fluidized carriers may be sand grains, granular activated carbon, diatomaceous earth, or other small solids that are resistant to abrasion. The upward velocity of the fluid must be sufficiency to maintain the carriers in suspension, and this depends upon the amount of biomass that is attached.

Rotating biological contactor (RBC)   The rotating biological contactor is another approach for a fixed-film reactor and, like the fluidized bed reactor, has good mixing and mass-transfer characteristics. Plastic media in a disk or drum has attached to the rotating shaft. As the shaft rotates, the portion of the contactor in contact with air absorbs oxygen, and the portion within the liquid absorbs contaminants to be oxidized. Wastewater can enter from one end of the RBC and travel perpendicular to the contactors, thus creating plug-flow character. Or, it can enter uniformly along the length of the reactor, and in this manner, can create a completely mixed system.

Aerators The equipment used for wastewater aeration is required for the biological process and also to provide mixing to keep solids suspended for more effective treatment. Although there are many types of aeration systems, the two basic methods of aerating wastewater are through mechanical surface aerators to entrain air into the wastewater by agitation or by introducing air or pure oxygen with submerged diffusers (see Figure 17.3). Aeration systems for conventional wastewater activated sludge plants typically account for 45% to 60% of a treatment facility's total energy use. Therefore, the impact of aeration systems on plant capital and operating costs is one measure of the importance of this unit operation to wastewater treatment.

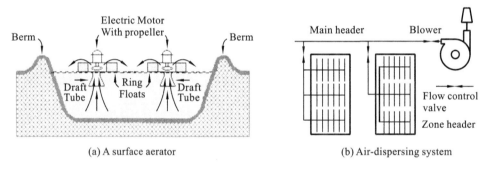

Figure 17.3 Various aerator types

Membrane Filtration Equipment Membrane filtration is the process that removes contaminants from a fluid by passage through a micro-porous membrane. Treatment processes that constitute membrane filtration include: microfiltration (MF), ultrafiltration (UF), nanofiltration (NF), and reverse osmosis (RO). MF is a type of filtration that uses special membranes with a pore size of approximately 0.03 to 10 microns. The size of the pores is what gives microfiltration its name. It is used to remove all particles that are larger than about 1 micrometer from a solution. Ultrafiltration is a method of removing very small particles from liquid. A membrane used with ultrafiltration usually has pores that are 0.01 to 0.001 microns in size. This is small enough to remove most bacteria, viruses, high molecular substances, and polymer-type molecules. The process is commonly used for treating drinking water in compliance with strict standards. NF membranes have a nominal pore size of approximately 0.001 microns. Pushing water through these smaller membrane pores requires a higher operating pressure than either MF or UF. Operating pressures are usually near 600 kPa (90 psi) and can be as high as 1,000 kPa (150 psi). These systems can remove virtually all cysts, bacteria, viruses, and humic materials and can be used to address pollution in water supplies, chemical spills, impurities in drinking water and other fluids human consume, and in desalination. RO can effectively remove all inorganic contaminants from water and can also remove radium, natural organic substances, pesticides, cysts, bacteria, and viruses. In RO system, pressure is exerted on the side with the concentrated solution to force the water molecules across the membrane to the fresh water side. Normally, the membrane material is manufactured from a synthetic polymer, although other forms, including ceramic and metallic "membranes", may be available. Currently, almost all membranes manufactured for drinking water production are made of polymeric material, since they are significantly less expensive than membranes constructed of other materials. The most common membrane configurations are hollow-fiber, spiral-wound, and cartridge (see Figure 17.4).

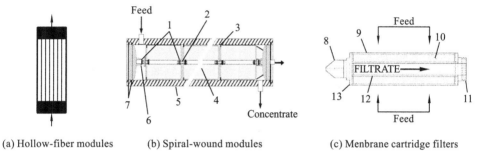

Figure 17.4 Various membrane configurations

1—O-rings; 2—interconnector; 3—brine seal; 4—spiral-wounded module;
5—pressure vessel; 6—end adapter; 7—head seal; 8—top end adapter; 9—outer case;
10—filter media; 11—bottom end adapter; 12—center core; 13—end cap

Most hollow-fiber modules used in drinking water treatment applications are manufactured to accommodate porous MF or UF membranes and designed to filter particulate matter. As the name suggests, these modules are comprised of hollow-fiber membranes, which are long and very narrow tubes. The fibers may be bundled in one of several different arrangements. In one common configuration used by many manufacturers, the fibers are bundled together longitudinally, potted in a resin on both ends, and encased in a pressure vessel that is included as a part of the hollow-fiber module. These modules are typically mounted vertically, although horizontal mounting may also be utilized.

Spiral-wound modules were developed as an efficient configuration for the use of semipermeable membranes to remove dissolved solids, and thus are most often associated with NF/RO processes. The basic unit of a spiral-wound module is a sandwich arrangement of flat membrane sheets called a "leaf" wound around a central perforated tube. One leaf consists of two membrane sheets placed back to back and separated by a fabric spacer called a permeate carrier. The layers of the leaf are glued along three edges, while the unglued edge is sealed around the perforated central tube. A single spiral-wound module 8 inches in diameter may contain up to approximately 20 leaves, each separated by a layer of plastic mesh called a spacer that serves as the feed water channel.

A representative diagram of membrane cartridge filter is shown in Figure 17.4 (c). Membrane cartridge filters are manufactured by placing flat sheet membrane media between a feed and filtrate support layer and pleating the assembly to increase the membrane surface area within the cartridge. The pleat pack assembly is then placed around a center core with a corresponding outer cage and subsequently sealed, via adhesive or thermal means, into its cartridge configuration. End adapters, typically designed with a double O-ring sealing mechanism, are attached to the filter to provide a positive seal with the filter housing.

New Words and Expressions

1. primary 首要的,初级的
2. activated sludge 活性污泥
3. tertiary 第三的,第三期的
4. aerator 曝气设备,增氧机
5. aerobic 耗氧的,需氧的
6. reverse osmosis 反渗透,反渗析
7. electrodialysis 电渗析
8. hollow-fiber module 中空纤维膜组件
9. ultrafiltration 超滤,超滤法
10. spiral-wound module 卷式膜组件
11. floatation 气浮,浮选
12. skim 撇去,撇除
13. cartridge 过滤筒,筒状物

Notes

(1) The tertiary system, or advanced treatment system, is becoming more prevalent to remove specific residual substances, trace organic materials, and to disinfect the water before discharge to a surface water stream or ocean outfall.

污水的三级处理或称高级处理系统普遍用于去除污水中的某些特殊残留污染物和微量元素,以及在污水被排入地表水系或入海口之前对其进行消毒。

(2) Suspended-growth systems include activated sludge, where the biomass is mixed with biodegradable organic compounds in the wastewater. Fixed-film or attached growth systems include trickling filters, bio-towers, and rotating biological contactors, etc, where the biomass grows on media and the wastewater passes over its surface.

悬浮生长净化系统中含有活性污泥,由活性污泥菌团与污水混合,对有机物进行降解。固定(或附着)膜净化系统包括滴流式生物滤器、生物塔和生物转盘(筒)等,生物膜附着在固体介质上,对流过其表面的污水进行净化。

(3) As the shaft rotates, the portion of the contactor in contact with air absorbs oxygen, and the portion within the liquid absorbs contaminants to be oxidized.

当生物转轴旋转时,暴露在空气中的盘体上的生物膜直接与空气接触并吸收氧气,浸没在污水中的盘体的生物膜对污染物进行分解净化。

(4) Although there are many types of aeration systems, the two basic methods of aerating wastewater are through mechanical surface aerators to entrain air into the wastewater by agitation or by introducing air or pure oxygen with submerged diffusers.

曝气设备类型有很多,目前污水处理中主要采用两种,即机械搅拌式表面曝气设备和水下空气或纯氧扩散设备。

(5) In one common configuration used by many manufacturers, the fibers are bundled together longitudinally, potted in a resin on both ends, and encased in a pressure vessel that is included as a part of the hollow-fiber module.

常用的中空纤维膜组件的制造方法是：首先将纤维膜沿其长度方向捆扎在一起，再将两端分别固定在树脂底座中，最后将捆扎和固定好的纤维膜安装在一个压力容器内，这样就组装完成了一套空纤维膜组件。

(6) Membrane cartridge filters are manufactured by placing flat sheet membrane media between a feed and filtrate support layer and pleating the assembly to increase the membrane surface area within the cartridge. The pleat pack assembly is then placed around a center core with a corresponding outer cage and subsequently sealed, via adhesive or thermal means, into its cartridge configuration.

筒式膜滤器的组装过程为：首先将平整的膜材料放置在分别与原液和滤出液接触的两个支撑层之间，再将三者褶皱叠起以增加滤膜的表面积；将褶皱的膜过滤材料安装在滤筒和外壳之间，用黏结剂或通过加热将模组件的两端密封好。

❖ Writing Training

如何写英文科技论文（二）

英文科技论文一般由摘要、序言、实验、结果与讨论、结论及参考文献等六个部分组成。

（一）摘要（abstract）

科技论文摘要通常包括以下几项内容：研究动机、研究方法、主要结果、简要结论。

摘要应该言简意赅，因此以上各项内容争取用一句话说明一项，每项最多不要超过三句话。通常在提交论文全文之前需要先提交摘要。摘要实际上就是一个小的 IMRaD 结构：为什么做这个研究？用了什么方法？取得了什么结果？结论是什么？这些问题逐一回答了，摘要就写完整了。

（二）引言（introduction）

引言看似简单，但并不容易写好。好的引言通常包括三部分内容：介绍研究课题（性质、范围等）；陈述关于该课题已有的主要研究成果；解释对该课题研究的特殊贡献，例如使用了什么新方法等。

对这一部分应该尽量使语言简练。好的起始句非常重要，因为引言应该吸引读者而不是让读者生厌。有许多论文以"It is widely accepted that … is important"开头，这样的起始句让读者还没进入正文就开始打哈欠。在引言这一部分可以简单介绍主要研究结果和结论，也可以不介绍结论而只介绍研究方法。读者读完引言之后看论文的其他部分不应该再有新的发现，因为读科技论文和读侦探小说不一样，读者希望在开头就知道结果。

（三）方法（methods）

方法部分的目的在于描述所用的材料/实验装置/实验方法/理论模型/计算方法。写好这部分的关键在于把握好"度"，即提供恰到好处的细节，避免过于简单或烦琐（太繁复或不必需的公式、推导可放入附录）。衡量标准是看文章所提供的细节是否足够以让感兴趣的专业读者重复你的实验或方法，在这一部分不需要汇报结果。

(四) 结论(results)

在结论部分只需要如实地汇报结果或数据即可,无须加入自己的解释——让结果和数据来表达研究结论。这一部分通常会包含图表。读者在阅读一篇论文时,往往在看完题目和摘要后就会浏览所有图表,有进一步兴趣时才会再读文章的其他部分,所以图表非常重要。它们不仅应该简明、清晰、准确,还应该完整,即每一个图表均应有详尽说明,使读者即使不看论文的文字部分也能够理解图表所要传达的信息。图表的顺序也很重要,它们应该体现行文的逻辑。有些作者习惯于将一系列图表陈列在一起,不在表头作解释,仅在文字中简单地进行介绍,期待读者自己去研究理解各个图表,这种做法是不可取的。

(五) 讨论(discussion)

讨论部分是论文的精髓所在,也是中国人普遍感到难写的部分,其内容可能包括:提炼原理,揭示关联,进行归纳,提出分析、模型或理论,解释结果(results)与作者的分析、模型或理论之间的联系,因为掺进了作者的观点和解释,这一部分在行文时需要注意语气,不可夸张,同时也要注意避免无关紧要或并不相关的内容。

(六) 结论(conclusions)

在论文的结论(conclusions)部分,作者应该总结并阐明论文的主要结果及其重要性,同时点明局限性或有所保留的地方。结论应该是水到渠成,不应有让读者感到惊奇的内容,通常也不应该引用文章其他部分未曾提及的文献。爱希比教授的"概念图"表明,结论实际上就是把结果部分和讨论部分的精要内容进行总结。结论可以分点陈述,简洁概括。

(七) 致谢(acknowledgement)

科学研究通常不是只靠一、两个人的力量就能完成的,需要多方面力量支持,协助或指导,特别是大型课题,更需联合作战,参与的人数很多。在论文结束时,应对在整个研究过程中曾给予自己帮助和支持的单位和个人表示谢意。尤其是对于参加部分研究工作却未署名的人,要肯定其贡献,予以致谢。

(八) 参考文献(references)

作者在论文之中,凡是引用他人的报告、论文等文献中的观点、数据、材料、成果等,都应按本论文中引用先后顺序排列,在文中标明参考文献的顺序号或引文作者姓名。每篇参考文献按篇名、作者、文献出处排列。列上参考文献的目的,不只是便于读者查阅原始资料,也便于自己进一步研究时参考。应该注意的是,凡列入的参考文献,作者都应详细阅读过,不能列入未曾阅读的文献。

对一般的科技文章来说,不一定需要上述文体结构都齐备,可以适当进行取舍。

❖ Reading Material

Equipment of Solid Waste Management

1. Introduction to solid waste

Solid wastes are all the wastes arising from human and animal activities that are normally solid and discharged as useless or unwanted. Because of their intrinsic properties, discarded waste materials are often able reused and may be considered a resource in another setting. Solid waste management may be defined as the discipline associated with the control of generation, storage, collection, transfer and transport, processing, and disposal of solid wastes in a manner that is in accord with the best principles of public health, economics, engineering, conservation, aesthetics, and other environmental considerations, and that is also responsive to public attitudes. In its scope, solid waste management includes all administrative, financial, legal, planning, and engineering functions involved in solutions to all problems of solid wastes.

The activities associated with the management of solid wastes have been group into the six functional elements: (1) waste generation; (2) waste handling and separation, storage, and processing at the source; (3) collection; (4) separation and processing and transformation of solid waste; (5) transfer and transport; and (6) disposal. Methods used for the processing and recovery municipal solid waste include size reduction, size separation, density separation, electric and magnetic field separation, densification (compaction) and material handling.

2. Size reduction of solid waste

Several types of size reduction units are in common use: (1) the hammermill, which is very effective with brittle materials; (2) the shear shredder, which uses two opposing counterrotating blades to cut ductile materials in a scissorlike action; and (3) the tub grinder, which is widely used in the processing of yard wastes.

Operationally, a hammermill shredder (see Figure 17.5(a)) is an impact device in which a number of hammers are fastened flexibly to an inner shaft or disk that is rotated at high speed (700 to 1200 rpm). Because of centrifugal force, the hammers extend radially from the centre shaft. As solid wastes enter the hammermill, they are hit with sufficient force to crush or tear them and with such a velocity that they do not adhere to the hammers. Wastes are further reduced in size by being struck against breaker plates or cutting bars fixed around the inner periphery of the inner chamber. The cutting action continuous until the material is of the size required and falls out of the bottom of the mill.

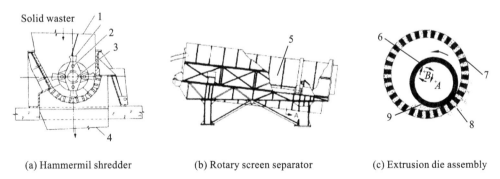

(a) Hammermil shredder   (b) Rotary screen separator   (c) Extrusion die assembly

Figure 17.5 Schematic diagram of typical size reduction units

1—hopper; 2—hammer; 3—shaft; 4—shredded solid waste; 5—drum screen;
6—press wheel; 7—extrusion die; 8—extruded cube; 9—solid waste

Size separation, or screening, involves the separation of materials by means of screening surfaces. The most common types of screens used for separation of solid waste are vibrating screens, rotary drum screens, and disc screens. The rotary drum screen is one of the most versatile types of screen for solid waste processing. It consists of a larger-diameter (typically 10 ft) screen, formed into a cylinder and rotating on a horizontal axis (see Figure 17.5(b)). Rotary drum screens have been used to protect shredders in refuse-derived fuels (RDF) production facilities (by removing oversized material) and to separate cardboard and paper in material recovery facilities.

Densification of solid waste is performed for several reasons, including reduction of storage requirement of recyclables, reduction of volume for shipping, and preparation of densified RDF. The type of equipment that available for the densification of solid wastes include stationary compactors; baling machines, which produce bales secured with wire or plastic ties; and cubing and pelleting machines, which produce woodlike cubes or pellets that are structurally stable because of chemical binding agents or heat treatment during the densification process. Cubing and pelleting machines operate on the similar principle. Waste paper or shredded RDF is extruded through extrusion dies with an eccentric rotating press wheel (see Figure 17.5(c)). The cubes or pellets are bonded together by heat cause by friction as the cubes or pellets are extruded. If kept dry, densified RDF can be stored for months without decomposition.

3. Thermal processing of solid waste

The thermal processing of solid waste, used both for volume reduction and energy recovery, is an important element in many wastes management systems. Thermal processing of solid waste can be defined as the conversion of solid wastes into gaseous, liquid, and solid conversion products, with the concurrent or subsequent release of heat energy. Combustion with exactly the amount of oxygen (or air) needed for complete combustion is known as stoichiometric combustion. Combustion with oxygen in excess of the stoichiometric requirements is termed excess-air combustion. Gasification is the partial combustion of solid waste under substoichiometric conditions to generate a combustible gas containing carbon monoxide, hydrogen, and gaseous hydrocarbons. Pyrolysis is the thermal processing of waste in the complete absence of oxygen.

In mass-fired combustion system, minimal processing is given to solid waste before it is placed in the charging hopper of the system. One of the most critical components of a mass-fired combustion system is the grate system(see Figure 17.6(a)). It serves several functions, including the movement of waste through the system, mix of the waste, and injection of combustion air. Many variation of grate are possible, based on reciprocating, rocking, or rotating elements.

A fluidized bed combustion(FBC) system consists of a vertical steel cylinder, usually refractory-lined, with a sand bed, a supporting grid plate, and air injection nozzles known as tuyeres(see Figure 17.6(b)). When air is forced up through the tuyeres, the bed fluidized and expands up to twice its resting volume. Solid fuels, such as coal or RDF, can be injected into the reactor below or above the level of the fluidized bed. The "boiling" action of the fluidized bed promotes turbulence and mixing and transfers heat to the fuel. In operation, auxiliary fuel (natural gas or fuel oils) is used to bring the bed up to operating temperature (1450 to

1750℉). After startup, auxiliary fuel is usually not needed; in fact, the bed remains hot up to 24 hours, allowing rapid restart without auxiliary fuel.

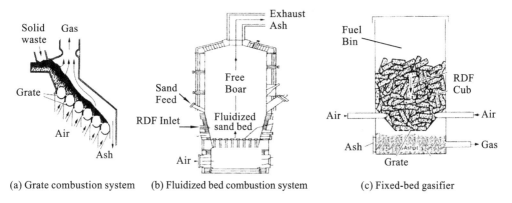

Figure 17.6 **Typical thermal processing units of solid waste**

There are five basic types of gasifiers: (1) vertical fixed-bed; (2) horizontal fixed-bed; (3) fluidized bed; (4) multiple hearth, and (5) rotary kiln. The vertical fixed bed gasifier has a number of advantages over the other types of gasifiers, including simplicity and relatively low capital costs. However, this type of reactor is more sensitive to the mechanical characteristics of the fuel; it requires a uniform, homogenous fuel, such as densified RDF. As shown in Figure 17.6(c), fuel flow through the gasifier is by gravity, with air and fuel flowing concurrently through the reactor. The end products of the process are primarily low-Btu gas and char. It is also possible to operate a vertical fixed bed reactor in a countercurrent flow mode, with air and gas moving upwards through the reactor.

New Words and Expressions

1. densification   压实,稠化
2. refractory-lined   耐火材料衬里的
3. baling   打包,捆扎
4. tuyere   风管口,鼓风口
5. pellet   颗粒,小球
6. hammermill shredder   锤式粉碎机
7. extrude   挤压
8. rotary drum screen   转筒式分级筛
9. stoichiometric   化学当量的
10. breaker plate   齿板
11. gasification   气化,气化作用
12. refuse-derived fuels (RDF)   废物基燃料
13. pyrolysis   高温分解,热解

Lesson 18 Remanufacturing Engineering

Text

As manufacturing generates in excess of 60% of non-hazardous waste every year, increasingly severe legislation demands a reduction in the environmental impacts of products and manufacturing processes. For example, producers have the responsibility to recover used products to reduce landfill. Such pressures, combined with severe competition due to global industrial activities, make companies to alter attitudes to product design. Companies must design products for longevity and ease of recovery of their materials at the end of life, and must consider the business potential of processing used products to harness the residual value in their components.

Remanufacturing, a process of bringing used products to a "like-new" functional state, can be both profitable and less harmful to the environment than conventional manufacturing as it reduces landfill and consumption of virgin material, energy, and specialized labor force used in production. Key barriers of remanufacturing include consumer acceptance, scarcity of remanufacturing tools and techniques and poor remanufacturability of many current products. These result from a lack of remanufacturing knowledge including ambiguity in its definition. Such terms as repair, reconditioning and remanufacturing are often used synonymously. Consequently, customers are unsure of the quality of remanufactured products and are wary of purchasing them. Also, designers may lack the knowledge to consider end-of-life issues such as remanufacturing in their work because design has traditionally focused on functionality and cost at the expense of environmental issue. In addition, the problems involved in organization transformations need investigation such as the interactions between innovations on technical, organizational and social development levels and timescales in these kinds of transformations. The decision-making processes used in manufacturing management, particularly in changing initiatives transformations need to be evaluated. Remanufacturing approach represents a fundamental change in the decision-making processes of most manufacturers. Historically, the selection of waste management methods was based on pure economic analyses of the quantifiable and measurable costs, and economic benefits. This approach ignores a very large number of qualitative factors affecting the selection of the appropriate technologies in decision-making. Remanufacturing, on the other hand, is a complex, multidisciplinary, and multifunctional activity method to determine potentially large numbers of waste minimization technologies available in the industry, for example, changes in the product, change in the input materials to the production process, changes in operating practices, and recycling. It requires the coordination of several designs and data-based activities, such as environmental impact analysis, data and database management, and design optimizations. There is now a consensus that decision-making is influenced by unpredictability, risk and uncertainty. For example, can the impacts of certain technologies be accurately calculated? How do you identify and calculate risks in different respects, e. g. economics, society an environment? And when certain practices succeed in one organization, why did they fail in another?

Remanufacturing engineering is a series of technical measures or engineering activities, which is based on the product life cycle theory(see Figure 18. 1), improves a waste product performance as the goal, is high quality, efficient, energy saving, materials, environmental protection as a criterion, has advanced technology and industrial production to repair, transform a waste product. In short, remanufacturing engineering is the industrialization of waste products by high-technology repair. It has proved that remanufacturing can develop and use the maximum value contained in waste products, and is the best form and preferred way of waste electrical and mechanical products reused, and is an important means to conserve resources. Remanufacturing of used mechanical and electrical products is an important measure for the development of circular economy and building a conservation-oriented society.

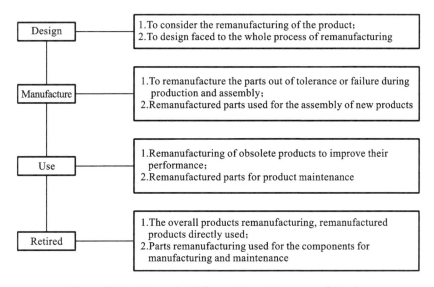

Figure 18. 1 **the product life cycle involving remanufacturing**

Components of mechanical and electrical equipments may have such failure modes as wear, deformation, fracture, erosion in use, and their accuracy, performance and productivity will reduce, which can lead to equipment failure, accident or even scrapped, so it is necessary to maintenance and repair them timely. In repair maintenance, all measures are aimed at the shortest time, the least cost to eliminate the failure effectively, and improve the effective utilization of the equipment. Repair process can make failure parts regenerate, and meet this purpose. There are common methods of repair, such as mechanical repair, welding, thermal spraying, electroplating, bonding, scraping, etc.. The following is the oxyacetylene flame spraying process to explain the repair process.

The process of oxyacetylene flame spraying

The schematic diagram of oxyacetylene flame spraying is seen in Figure 18.2. The process of oxyacetylene flame spraying includes surface treatment, preheating, spraying base powder, spraying powder of the working coating, machining of sprayed coating.

In addition to copper and tungsten-based materials, all common steel, stainless steel, hardened alloy steel, nitride steel, nickel, chromium alloys, cast iron can be sprayed.

1) The preparation before spraying

The preparation before spraying involves such processes as workpiece cleaning, surface preparation, surface roughening, and pre-heating.

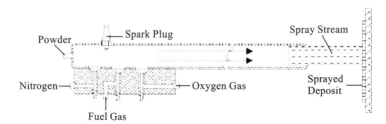

Figure 18.2 the schematic diagram of oxyacetylene flame spraying

The main object of workpiece cleaning is regions of workpiece to be sprayed, oil, corrosion, and oxidation near them. Flame baking is used to skim for some materials to guarantee the quality of combination.

The purpose of surface preparation is to remove fatigue layer, carburized layer, coating of the workpiece surface, and surface damage, correct uneven wear surface, reserve coating thickness, which determines pre-processing capacity. There are such common methods of surface preparation as turning and grinding.

Surface roughening is the roughening treatment for the surface to be sprayed to improve the bonding strength between sprayed coatings and substrate. There are such common methods of surface roughening as sandblasting and EDM roughening. Of course, such machining methods may be used as turning, grinding, and knurled. The roughening treatment as turning is to process the thread which pitch ranges from 0.3 mm to 0.7 mm, and depth from 0.3 mm to 0.5 mm.

The purpose of preheat is to remove water of the surface adsorption, reduce shrinkage stress during the cooling, and improve the bonding strength.

2) Spraying binding coating

Binding coating should be immediately sprayed on the workpiece after pretreatment to improve the bonding strength between working coatings and substrate. It is particularly applicable in the case of thin workpiece, or easy deformation when sandblasted.

The thickness of binding coating ranges from 0.1 mm to 0.15 mm, and the distance of spraying from 180 mm to 200 mm.

3) Spraying working coating

The working coating should be immediately sprayed on the workpiece after spraying binding coating. The quality of working coating depends on the powder feed rate and spray distance. The powder feed rate is large to increase raw powder, and reduce the quality of coating. The powder feed rate is small to lower productivity. The distance of spraying is near to have no sufficient time to heat powder, which lead to high temperature of workpiece. The distance of spraying is far to reduce the speed and temperature of alloy powder. The linear velocity of the workpiece surface ranges from 20 m to 30 m per minute. The powder injection direction is perpendicular to the surface of spraying during spraying.

4) Treatment after spraying

It is cooling slowly after spraying. Anticorrosive fluid is sprayed to keep coating from corrosion because sprayed coatings porous. Under normal circumstances paint and epoxy resin is sprayed the surface of coating.

When the dimensional accuracy and surface roughness of the coating can not meet the requirements, turning or grinding can be used for finishing.

New Words and Expressions

1. remanufacture　再制造
2. legislation　法律,法规
3. landfill　垃圾填埋
4. harness　利用,管理
5. mechanical repair　机械修复
6. thermal spraying　热喷涂
7. electroplating　电镀
8. bonding　胶结
9. scraping　刮研
10. oxyacetylene flame spraying　氧乙炔焰喷涂
11. surface treatment　表面处理
12. preheating　预热
13. base powder　基体粉末
14. coating　涂层
15. working coating　工作层
16. surface roughening　表面硬化
17. substrate　基体
18. sandblast　喷砂处理
19. EDM　电火花
20. epoxy resin　环氧树脂

Notes

(1) Companies must design products for longevity and ease of recovery of their materials at the end of life, and must consider the business potential of processing used products to harness the residual value in their components.

公司必须设计寿命长、在报废时材料容易回收的产品,必须考虑废旧产品加工的商业潜力,以挖掘零部件的剩余价值。

(2) Remanufacturing, a process of bringing used products to a "like-new" functional state, can be both profitable and less harmful to the environment than conventional

manufacturing as it reduces landfill and consumption of virgin material, energy, and specialized labor force used in production.

再制造是将使用过的产品在其功能上"整旧如新"的过程,此过程不仅可获利,而且比常规过程对环境危害作用小,因为它能减少垃圾填埋和生产过程中原材料、能源及具有专门技术的劳动力的消耗量。

(3) Remanufacturing engineering is a series of technical measures or engineering activities, which is based on the product life cycle theory, improves a waste product performance as the goal, is high quality, efficient, energy saving, materials, environmental protection as a criterion, has advanced technology and industrial production to repair, transform a waste product.

再制造工程是以产品全寿命周期理论为指导,以废旧产品性能跨越式提升为目标,以优质、高效、节能、节材、环保为准则,以先进技术和产业化生产为手段,来修复、改造废旧产品的一系列技术措施或工程活动的总称。

(4) Components of mechanical and electrical equipments may have such failure modes as wear, deformation, fracture, erosion in use, and their accuracy, performance and productivity will reduce, which can lead to equipment failure, accident or even scrapped, so it is necessary to maintenance and repair them timely.

在机电设备的使用过程中,零部件会产生磨损、变形、断裂、蚀损等形式的失效,设备的精度、性能和生产率将下降,从而导致设备发生故障、事故甚至报废,因而需要及时地对其进行维护和修理。

❖ Writing Training

认识科技论文的英文摘要

摘要(abstract)作为对研究论文正文的精练概括,有利于读者在最短的时间内了解全文内容。随着国际检索系统的出现,摘要逐渐成为一种信息高度密集的相对独立文体,为人们在浩如烟海的文献中寻找所需要的信息提供了便利。

随着二次文献数据库的普及以及全球科学技术界对科技信息日益增长的需求和重视,论文摘要的受关注率比论文本身要大数十倍甚至数百倍。为此,一篇论文能否得到重视,能否把科研成果准确地传播出去,能否被更多重要的数据库收录,与摘要的内容和质量的好坏直接相关。

现在多数大学要求学生的毕业论文应有英文摘要(含英文题目及关键词),所以有必要了解摘要的写作。

• 定义与分类

摘要置于正文前面,是对文献的内容不加任何解释(interpretation)和评论(evaluation)的简要而准确的(concise and precise)表达(description)。

注　Abstract, Synopsis, Summary 的区别:
Abstract　摘要,文摘　置于文前。
Synopsis　梗概,用于电影(movie)、故事(story)、小说(fiction)等。
Summary　概述,置于文尾。

- 分类

摘要一般分为两类：信息性摘要和说明性或指示性摘要，目前绝大部分的科技期刊和会议论文都要求作者提供信息性摘要。

1) 信息性摘要(informative abstract)

主要强调尽量多而完整地报道原文献中的具体内容，特别是研究目的，研究问题，研究方法和手段，主要论点和发现，得出的结论以及建议、措施等。它包括文章的主旨和数据等，适用于学位论文(dissertation)、学术刊物论文(journal paper)、学术会议论文(conference paper)、学术海报(poster)。

2) 说明性摘要(description abstract or indicative)

提供主要内容(problem or issue)，但不介绍具体内容(content)，适用于讨论性文章等。

- 作用及特点

1) 作用

便于搜索、查阅、浏览各种文献。

摘要独立于正文，通常收录于相应学科的摘要检索类数据库或专刊内，撰写好摘要有利于论文被数据库收录和他人引用。

摘要的目的是为读者提供关于文献内容的有用信息，即论文所包含的主要概念和所讨论的主要问题。读者从摘要中可获知作者的主要研究活动、研究方法和主要研究结果及结论。它可以帮助读者判断此论文对自己的研究工作是否有用，是否有必要获取全文，为科研人员、科技情报人员及计算机检索提供方便。

2) 特点

(1) 独立性：摘要包括使读者理解原文献的基本要素，可离开原文独立存在。

(2) 概括性：摘要把一篇文章的精华部分以精炼的文字、极短的篇幅概括出来，成为浓缩的信息。

(3) 客观性：摘要是对文章不加评论、解释的客观报道。

❖ Reading Material

Surface Engineering

Surface Engineering is one of the key technologies of remanufacturing engineering. Remanufacturing engineering is the carrier of the practical applications of surface engineering. Surface engineering technology can improve the resource utilization of remanufactured products.

The process of surface engineering, or surface treatments, tailor the surfaces of engineering materials to: (1) control friction and wear, (2) improve corrosion resistance, (3) change physical property, e. g. , conductivity, resistivity, and reflection, (4) alter dimension, (5) vary appearance, e. g. , color and roughness, (6) reduce cost. Common surface treatments can be divided into two major categories: treatments that cover the surfaces and treatments that alter the surfaces.

1. Covering the surfaces

The treatment that cover the surfaces include organic coatings and inorganic coatings.

The inorganic coatings perform electroplating, conversion coatings, thermal spraying, hot dipping, furnace fusing, or coat thin films, glass, ceramic on the surface of the materials.

Electroplating is an electrochemical process by which metal is deposited on a substrate by passing a current through the path.

Usually there is an anode (positively charged electrode), which is the source of the material to be deposited; the electrochemistry which is the medium through which metal ions are exchanged and transferred to the substrate to be coated; and a cathode (negatively charged electrode) which is the substrate to be coated.

Plating is done in a plating bath which is usually a non-metallic tank (usually plastic). The tank is filled with electrolyte which has the metal, to be plated, in ionic form.

The anode is connected to the positive terminal of the power supply. The anode is usually the metal to be plated (assuming that the metal will corrode in the electrolyte). For ease of operation, the metal is in the form of nuggets and placed in an inert metal basket made out non-corroding metal (such as titanium or stainless steel).

The cathode is the workpiece, the substrate to be plated. This is connected to the negative terminal of the power supply. The power supply is well regulated to minimize ripples as well to deliver a steady predictable current, under varying loads such as those found in plating tanks.

As the current is applied, positive metal ions from the solution are attracted to the negatively charged cathode and deposit on tire cathode. As a replenishment for these deposited ions, the metal from the anode is dissolved and goes into the solution and balance the ionic potential.

Thermal spraying process. Thermal sprayed metal coatings are depositions of metal which has been melted immediately prior to projection onto the substrate. The metals used and the application systems used vary but most applications result in thin coatings applied to surfaces requiring improvement to their corrosion or abrasion resistance properties.

Thermal spraying is a generic term for a broad class of related processes in which molten droplets of metals, ceramics, glasses, and/or polymers are sprayed onto a surface to produce a coating, to form a free-standing near-net-shape, or to create an engineered material with unique properties.

In principle, any material with a stable molten phase can be thermally sprayed, and a wide range of pure and composite materials are routinely sprayed for both research and industrial applications. Deposition rates are very high in comparison to alternative coating technologies. Deposit thickness of 0.1 mm to 1 mm is common, and thickness greater than 1 cm can be achieved with some materials.

2. Altering the Surfaces

The treatments that alter the surfaces include hardening treatments, high-energy processes and special treatments.

High-energy processes are relatively new surface treatment method. They can alter the properties of surfaces without changing the dimension of the surfaces. Common high-energy processes, including electron beam treatment, ion implantation, and laser beam treatment, are briefly discussed as follows:

Electron beam treatment   Electron beam treatment alters the surface properties by rapid heating —using electron beam and rapid cooling—in the order of 10^6 ℃/s in a very shallow region, 100 μm, near the surface. This technique can also be used in hard-facing to produce "surface alloys".

Ion implantation   Ion implantation uses electron beam or plasma to impinge gas atoms to ions with sufficient energy, and embed these ions into atomic lattice of the substrate, accelerated by magnetic coils in a vacuum chamber. Tire mismatch between ion implant and the surface of a metal creates atomic defects that harden the surface.

Laser beam treatment   Similar to electron treatment, laser beam treatment alters the surface properties by rapid heating and rapid heating and rapid cooling in a very shallow region near the surface. It can also be used in hard-facing to produce "surface alloys".

The results of high-energy processes are not well known or very well controlled. But the preliminary results look promising. Further development is needed in high-energy processes, especially in implant dosages and treatment methods.

New Words and Expressions

1. surface engineering   表面工程
2. surface treatment   表面处理
3. corrosion resistance   耐(腐)蚀性，耐蚀力,抗腐(蚀)性
4. conductivity   电导率
5. resistivity   电阻率
6. reflection   反射率
7. coating   涂层
8. electroplating   电镀
9. conversion coating   涂层转换
10. thermal spraying   热喷涂
11. hot dipping   热浸
12. furnace fusing   炉内重熔
13. electron beam treatment   电子束处理
14. ion implantation   离子注入
15. plasma   等离子
16. impinge   冲击,影响

Lesson 19 Mining Metallurgy Equipment

Text

Mineral processing is the first process that most ores undergo after mining in order to provide a more concentrated material for the procedures of extractive metallurgy. The primary operations are comminution and concentration, but there are other important operations in a modern mineral processing plant, including sampling and analysis and dewatering.

After mining, large pieces of the ore feed are broken through crushing and/or grinding in order to obtain particles small enough where each particle is either mostly valuable or mostly waste. The next stage of treatment is concentration. The great importance of concentration processes in modern metallurgy results from the desire to increase the efficiency of metallurgical production, and also from the use of lower-grade ores required by the growth of metal smelting. The direct metallurgical treatment of such ores without concentration is usually not economically feasible and is sometimes even impossible. Flotation, gravity, magnetic, and electrical methods of concentration are most widespread. In one such method, the ore is crushed and placed in a machine where, by shaking, the heavier particles containing the metal are separated from the lighter rock particles by gravity. Flotation processes are used to treat more than 90 percent of the ore of nonferrous and rare metals. In certain cases (as when gold, silver, or occasionally copper occur "free", i. e. , uncombined chemically in sand or rock), mechanical or ore dressing methods alone are sufficient to obtain relatively pure metal. Waste material is washed away or separated by screening and gravity; the concentrated ore is then treated by various chemical processes.

Following concentration by mineral processing, metallic minerals are subjected to extractive metallurgy, in which their metallic elements are extracted from chemical compound form and refined of impurities. Since almost all the metals are found combined with other elements in nature, chemical reactions are required to set them free. These chemical processes are classified as pyrometallurgy, electrometallurgy, and hydrometallurgy.

Pyrometallurgy, or the use of heat for the treatment of an ore, includes roasting and smelting. Metallic compounds are frequently rather complex mixtures, and they are not often types that permit extraction of the metal by simple, economical processes. Consequently, before extractive metallurgy can effect the separation of metallic elements from the other constituents of a compound, it must often convert the compound into a type that can be more readily treated. The preliminary treatment that is commonly used to do this is roasting in which compounds are converted at temperatures just below their melting points. There are several different types of roast, each one intended to produce a specific reaction and to yield a roasted product (or calcine) suitable for the particular processing operation to follow. A sulfide ore is commonly roasted, i. e. , heated in air. The metal of the ore combines with oxygen of the air to form an oxide, and the sulfur of the ore also combines with oxygen to form sulfur dioxide, which, being a gas, passes off. The metallic oxide is then treated with a reducing agent. The roasting processes can be carried out in specialized roasters. Fluidized-bed roasters (see Figure 19. 1) have found wide acceptance because of their high capacity and efficiency. They can be used for oxidizing, sulfatizing, and volatilizing roasts. The roaster is a refractory-lined, upright cylindrical steel shell with a grate bottom through which air is blown in sufficient volume to keep fine, solid feed particles in suspension and give excellent gas-solid contact. The ore feed can be introduced dry or as a water suspension through a downpipe into the turbulent layer zone of the roaster. Discharge of the roasted calcines is through a side overflow pipe.

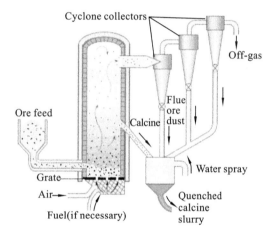

Figure 19. 1 Fluidized-bed roaster

Smelting is a process that liberates the metallic element from its compound as an impure molten metal and separates it from the waste rock part of the charge by heating beyond the melting point. If the ore is an oxide, it is heated with a reducing agent, such as carbon in the form of coke or coal; the oxygen of the ore combines with the carbon and is removed in carbon dioxide, a gas. The waste material in the ore is called gangue; it is removed by means of a substance called a flux which, when heated, combines with it to form a molten mass called slag. Being lighter than the metal, the slag floats on it and can be skimmed or drawn off. The flux used depends upon the chemical nature of the ore; limestone is usually employed with a siliceous gangue.

Electrometallurgy includes the preparation of certain active metals, such as aluminum, calcium, barium, magnesium, potassium, and sodium, by electrolysis: a fused compound of the metal, commonly the chloride, is subjected to an electric current, the metal collecting at the cathode.

Hydrometallurgy consists of such operations as leaching, in which metallic compounds are selectively dissolved from an ore by an aqueous solvent, and electrowinning, in which metallic ions are deposited onto an electrode by an electric current passed through the solution. For example, certain copper oxide and carbonate ores are treated with dilute sulfuric acid, forming water-soluble copper sulfate. The metal is recovered by electrolysis of the solution.

Extraction is often followed by refining, in which the level of impurities is brought lower or controlled by pyrometallurgical, electrolytic, or chemical means. Pyrometallurgical refining usually consists of the oxidizing of impurities in a high-temperature liquid bath. Electrolysis is the dissolving of metal from one electrode of an electrolytic cell and its deposition in a purer form onto the other electrode. Chemical refining involves either the condensation of metal from a vapour or the selective precipitation of metal from an aqueous solution.

New Words and Expressions

1. ore   矿石
2. extractive metallurgy   提取冶金(学)
3. comminution   粉碎,研磨
4. concentration   精选
5. dewatering   脱水,除去水分
6. flotation   浮选
7. ore dressing   选矿
8. screening   筛选,过筛,屏蔽
9. compound   化合物
10. pyrometallurgy   火法冶金(学)
11. electrometallurgy   电冶金(学)
12. hydrometallurgy   湿法冶金(学)
13. roasting   焙烧,煅烧
14. smelting   熔炼,冶炼
15. reducing agent   还原剂
16. sulfatizing   硫酸盐化,硫酸化
17. volatilizing   挥发
18. refractory   耐火材料
19. calcine   煅烧产物,焙烧矿,焙砂
20. flux   溶剂,造渣
21. slag   炉渣,夹杂
22. cathode   阴极
23. leaching   浸出,溶出
24. aqueous solvent   水溶剂
25. electrowinning   电解沉积,电解冶金法
26. fluidized-bed roaster   沸腾炉

Notes

(1) The great importance of concentration processes in modern metallurgy results from the desire to increase the efficiency of metallurgical production, and also from the use of lower-grade ores required by the growth of metal smelting. The direct metallurgical treatment of such ores without concentration is usually not economically feasible and is sometimes even impossible.

在现代冶金中,精选工序是非常重要的,这是由于人们希望提高冶金生产的效率,也由于金属冶炼业的发展要求利用更低等级的矿石。不经过精选而直接对这些矿石进行冶金处理在经济上通常是不可行的,有时候甚至是不可能的。

(2) Consequently, before extractive metallurgy can effect the separation of metallic

elements from the other constituents of a compound, it must often convert the compound into a type that can be more readily treated.

因此,进行提取冶金时,在把金属元素与化合物的其他成分分离开来之前,通常必须先把化合物转变成可以更容易被处理的类型。

(3) The roaster is a refractory-lined, upright cylindrical steel shell with a grate bottom through which air is blown in sufficient volume to keep fine, solid feed particles in suspension and give excellent gas-solid contact.

焙烧炉的内衬采用耐火材料,有竖直的圆柱钢壳。炉底带栅格,可使足量的空气通过而进入炉内,使细固体给料颗粒保持悬浮状态,从而使气体与固体接触良好。

(4) Hydrometallurgy consists of such operations as leaching, in which metallic compounds are selectively dissolved from an ore by an aqueous solvent, and electrowinning, in which metallic ions are deposited onto an electrode by an electric current passed through the solution.

湿法冶金由浸出操作和电解冶金组成。在浸出操作中,利用水溶剂将金属化合物有选择性地从矿石中溶解出来;在电解操作中,在溶液中通以电流,使金属离子沉积在电极上。

❖ Writing Training

科技论文的英文摘要的写作与科技论文的标题

一、摘要的写作

• **基本内容及形式**

(1) 基本内容:包括研究工作的目的、方法、结果、结论及建议等。重点是结果和结论,即应突出论文的创造性成果和新见解。

(2) 摘要长度:常用 100~250 个词;学位论文、长篇报告 500 词以下,仅限于一页。

(3) 文体:用主题句开头,阐明原文主题,但注意第一句不得与原文题名完全重复,以免检索系统收录后有关人员用计算机检索时出现差错。

(4) 段落:一般不分段,对于学位论文等,摘要较长时可分段。

(5) 句子:用完整的句子,不用电报式文字、短语,注意前后应连贯。

(6) 用词:动词尽量用主动语态,有时亦用被动语态。

(7) 人称:常用第三人称,如必须可用第一人称;尽量用正式的书面语用词汇而不用口语化的词汇,尽量用简单的词汇而不用复杂的、生僻的词汇。

(8) 术语:避免用人们不翻译的术语、首字母缩写词及缩略语符号等。

(9) 顺序:摘要应在写完文章后撰写,以便将重要内容简洁地表述出来。

• **时态问题**

图表介绍、公式说明、试验结果、方法与真理描述等用"现在时";

试验经过、试验过程、过去做的研究工作、特殊结论及推论用"过去时";

叙述从某一时间开始,对现在有直接关系并影响的用"现在完成时";

今后的研究及打算、预期的结果、数学公式推演结果等用"将来时"。

二、标题的基本特点和要求

- 特点

标题用词应简明扼要,一般词组较多,且多为名词性词组,如"中心词＋修饰语"的形式,一般不用句子,如"Methods to Reduce Steel Wear in Grinding Mills"。也可用冒号以突出题旨,如"Advanced Vocabulary Learning, the Problem of Collocation Vocabulary: Learning to Be Imprecise"。

- 题目书写要求

（1）除冠词、连词及五个字母以下的介词外,其余所有的实词的首字母均应大写,当介词、连词在题目开头时,单词的第一个字母也应大写（有些刊物规定,题目的所有字母都大写）。

（2）标题字数宜在 20 字以内。如需分解,则第二行第一个单词的首字母应大写。

（3）除了破折号,一般不用（或很少用）标点符号。

（4）每个用词要考虑到有助于编制题录、索引和关键字等。

（5）力求简、明、短,不用完整的句子,多用名词性词组。

例如：

Abstract: The concept and connotation of, and the main measures for "Green manufacturing" are briefly discussed from the viewpoint of sustainable development. The issue is more closely dealt with in regard to the steel industry. It is pointed out that adequate environment protection in a "green" steel plant does not merely mean a tolerate disposal of pollutants emitted from its operation units, but rather the effective implementation of a strategy. The formation of any polluting agents in any part of this plant includes proper choice and control of raw materials, and an untiring endeavor effort to optimize the complete manufacturing process of the whole steel plant.

❖ Reading Material

Blast Furnace

A blast furnace (see Figure 19.2) is a type of metallurgical furnace used for smelting to produce industrial metals. Blast furnaces differ in construction. The one used in the production of iron consists of a chimneylike structure (usually 80~100 ft/24~30 m high) made of iron or steel and lined with firebrick. It is narrow at the top, increasing in diameter downward, but narrowing again suddenly almost at the bottom, to form the hearth. There the fine molten products are caught. The furnace is fed from the top with a charge of definite quantities of ore, coke, and a flux, mostly limestone. The raw materials are brought to the top of the blast furnace via a skip car powered by winches or conveyor belts. There are different ways in which the raw materials are charged into the blast furnace. Some blast furnaces use a "double bell" system where two "bells" are used to control the entry of raw material into the blast furnace. The purpose of the two bells is to minimize the loss of hot gases in the blast furnace. A more recent design is to use a "bell-less" system. These systems use multiple hoppers to contain each raw material, which is then discharged into the blast furnace through valves. These valves are more accurate at controlling how much of each constituent is added, thereby increasing the efficiency of the furnace. Some of these bell-less systems also implement a chute in order to precisely control where the charge is placed.

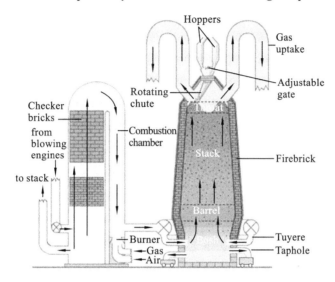

Figure 19.2 **Blast furnace**

Preheated compressed air is introduced at the bottom through pipes (tuyeres) entering just above the hearth. The air passes upward through the charge. The coke is oxidized to carbon monoxide at the high temperature. The carbon monoxide then reduces the iron ore to metallic iron and, taking on oxygen, reverts to carbon dioxide. This gas, together with unused carbon monoxide, nitrogen, and other constituents of the air originally introduced, is led off through four uptakes from the top of the furnace and, being still at a high temperature, is employed to heat the stoves into which fresh air for the process is brought. As the operation proceeds, the mass in the furnace becomes molten and descends into the hearth. The molten iron which picks up considerable quantities of carbon, manganese, phosphorus, sulfur, and silicon sinks to the bottom; The gangue(mostly silica) of the iron ore and the ash in the coke combine with the limestone to form the blast-furnace slag. Slag, being lighter, floats on top. The slag is drained through a pipe in the upper portion of the hearth. The iron is tapped from below and run into sand molds to harden, which is known as pig iron.

The "pig iron" has a relatively high carbon content of around $4\% \sim 5\%$, making it very brittle, and of limited immediate commercial use. The blast furnace produces pig iron for one of three applications: (1) the huge majority passes to the steelmaking process for refining; (2) pig iron is used in foundries for making cast iron; and (3) ferroalloys, which contain a considerable percentage of another metallic element, are used as addition agents in steelmaking.

The blast furnace remains an important part of modern iron production. Modern furnaces are highly efficient, including Cowper stoves to preheat the blast air and employ recovery systems to extract the heat from the hot gases exiting the furnace. Competition in industry drives higher production rates. Efforts to increase production rates have led to the addition of pure oxygen and steam and the sizing of ore to obtain better gas-solid contact. Flux and ore are sometimes combined into pellets. The largest blast furnaces have a volume around 5580 m^3 (190,000 cu ft) and can produce around 80,000 tonnes of iron per week.

Although the efficiency of blast furnaces is constantly evolving, the chemical process inside the blast furnace remains the same. According to the American Iron and Steel Institute:"Blast furnaces will survive into the next millennium because the larger, efficient furnaces can produce hot metal at costs competitive with other iron making technologies." One of the biggest drawbacks of the blast furnaces is the inevitable carbon dioxide production as iron is reduced from iron oxides by carbon and there is no economical substitute—steelmaking is one of the unavoidable industrial contributors of the CO_2 emissions in the world. The challenge set by the greenhouse gas emissions of the blast furnace is being addressed in an on-going European Program called ULCOS (Ultra Low CO_2 Steelmaking). Several new process routes have been proposed and investigated in depth to cut specific emissions (CO_2 per ton of steel) by at least 50%. Some rely on the capture and further storage (CCS) of CO_2, while others choose decarbonizing iron and steel production, by turning to hydrogen, electricity and biomass.

New Words and Expressions

1. blast furnace 高炉,鼓风炉
2. hopper 加料斗,给料斗
3. firebrick 耐火砖
4. throat 炉喉
5. hearth 炉缸,炉膛
6. dioxide 二氧化物
7. barrel 炉腰
8. limestone 石灰石
9. bosh 炉腹
10. combustion chamber 燃烧室
11. pig iron 生铁
12. tuyere 鼓风口
13. charge 炉料
14. taphole 出铁口,出渣口
15. uptake 垂直管道,上气道
16. drain 排水,流出
17. bell 料钟
18. checker brick 格子形耐火砖
19. winch 绞车
20. rotating chute 旋转斜槽
21. gangue 脉石
22. refining 精炼
23. monoxide 一氧化物
24. conveyor 运输机,传送带
25. American Iron and Steel Institute 美国钢铁学会
26. Cowper stove 考珀(鼓风)炉
27. pellet 粒状产品,球团矿
28. constituent 组元,组分,成分

Lesson 20 Sustainable Product Design

Text

1. Introduction

Sustainable design (also called green design or eco-design) is an approach to design a product with special consideration for the environmental impacts of the product during its whole lifecycle (see Figure 20.1). The application of sustainable design involves a particular framework for consideration environmental issues, the application of relevant analysis and synthesis methods, and a challenge to traditional procedures for design and manufacturing.

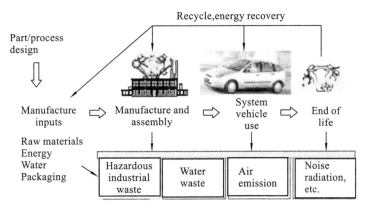

Figure 20.1 Life cycle of an automobile

In many past situations, environmental effects were ignored during the design stage for new products and processes. Hazardous wasted were dumped in the most convenient fashion possible, ignoring possible environmental damages. In efficient energy use results in high operation costs. Waste was common in material production, manufacturing and distribution. Consumers cast aside products, usually with only minimal remanufacturing or recycling. Recognition of theses problems inspired environmental engineering applications to clean up past pollution (called remediation) and ongoing wastes streams (called waste treatment). Clean up are still needed in many cases. But design changes can often be more effective at reducing environmental burdens and more efficient at reducing costs than traditional "end-of-the-pipe" clean up strategies. Some examples of such practices include:

Solvent substitution in which single use of a toxic solvent is replaced with a more benign alternatives such as biodegradable solvents or non-toxic solvents. Water based solvents are preferable to organic based solvents.

Technology changes such as more energy efficient semi-conductors or motor vehicle engines. For example, the "Energy Star" program specifies maximum energy consumption standards for computers, printers and other electronic devices. Products in compliance can be labeled with the "Energy Star". Similarly, "Green Lights" is a program that seeks more light from less electricity.

Recycling of toxic wastes can avoid dissipation of the materials into the environment and avoid new production. For example, rechargeable nickel-cadmium batteries can be recycled to recover both cadmium and nickel for other uses.

The challenge of sustainable design is to alter conventional design and manufacturing procedures to incorporate environmental considerations systematically and effectively. This requires changes in these existing procedures. Changing design procedures is particularly difficult because designers face many conflicting objectives, uncertainties, and a work environment demanding speed and cost effectiveness. Environmental concerns must be introduced in practical and meaningful fashions into these complicated design processes. Fundamental knowledge must be developed and new innovative technologies established to meet this need. Engineers must move beyond their traditional considerations of functionality, cost, performance, and time-to-market, to consider also sustainability. Engineers must begin thinking in terms of minimizing energy consumption, waste-free manufacturing processes, reduced material utilization, and resource recovery following the end of product use — all under the umbrella of a total life cycle view (see Figure 20.2). Of course, all those must be done with involvement of stakeholders, and the development of innovative technologies, tools, and methods.

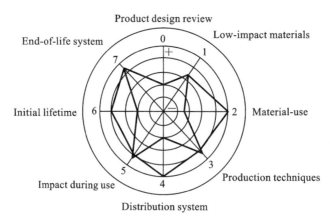

Figure 20.2 Eco-design strategy wheel

2. Objectives for sustainable design

There is no widespread consensus or agreement on the particular goals to be pursued by sustainable design. Some argue that sustainable design should be pursued solely to reduce costs. In this view, any waste from a process or product is an opportunity. Others focus on particular strategies, such as recycling to save raw materials, and develop goals specifically for these strategies. Another approach is to direct all attention to a particular environmental problem, such as global warming, or a particular media, such as air pollution, and ignore other environmental effects. Each of these approaches is flawed.

The social goals forsustainable design relate to ensuring a sustainable future for our society, in regard to both resources and ecological health. We can advance three general goals for sustainable design in pursuit of a sustainable future:
- Reduce or minimize the use of non-renewableresources;
- Manage renewable resources to insure sustainability;
- Reduce, with the ultimate goal of eliminating, toxic and otherwise harmful emissions to the environment, including emissions contributing to global warming.

The objective of sustainable design is to pursue these goals in the most cost-effective fashion.

3. Sustainable design strategies

A central concept insustainable design is the notion that the systematic effects of design decisions should be considered. In designing a new product, the environmental burdens associated with material supply, manufacture, use and disposal may all be relevant. Some strategies for sustainable design include:

Life cycle assessment (LCA) is a technique to assess environmental impacts associated with all the stages of a product's life from cradle-to-grave (see Figure 20.3 i. e., from raw material extraction through materials processing, manufacture, distribution, use, repair and maintenance, and disposal or recycling). LCA can help avoid a narrow outlook on environmental concerns by:

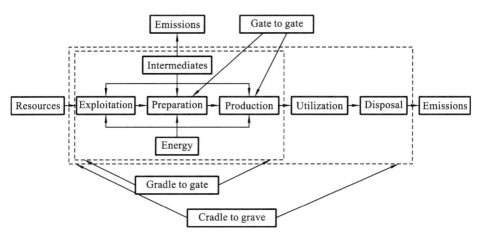

Figure 20.3 Boundaries of a product system

- Compiling an inventory of relevant energy and material inputs and environmental releases;
- Evaluating the potential impacts associated with identified inputs and releases;
- Interpreting the results help you to make a more informed decision.

Material flows and cycles is a technique for tracing material use and location over time (see Figure 20.4). For example, steel is routinely recovered from products such as automobiles, melted and re-used in a closed recycle loop. Some materials are disposed of into landfills, although the material may sometime in the future be recovered and re-used. In recycling materials, there is a distinction between closed-loop (re-use for the same function) and open loop (re-use in a different function, typically with lower quality requirements). In tracing materials flows, it is important to be clear about the boundaries of analysis and the uncertainty of mass measurements.

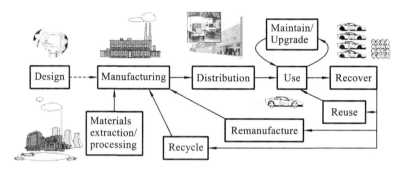

Figure 20.4 Material flow in a product life cycle for sustainability

Three R's in source management: The 3R's of reduce, reuse and recycle have been considered to be a base of environmental awareness and will lead to savings in materials and energy which will benefit the environment. Source reduction refers to any change in the design, manufacture, purchase, or use of materials or products (including packaging) to reduce their amount or toxicity before they become municipal solid waste. Reuse is to use an item more than once. This includes conventional reuse where the item is used again for the same function and new-life reuse where it is used for a different function. Recycling turns materials that would otherwise become waste into valuable resources. Collecting used bottles, cans, and newspapers and taking them to the curb or to a collection facility is just the first in a series of steps that generates a host of financial, environmental, and social returns. Some of these benefits accrue locally as well as globally.

4. Sustainable Design Methods

In designing a new product, the environmental burdens associated with material supply, manufacture, use and disposal may all be relevant. Designing and manufacturing sustainable products require appropriate knowledge, tools, production methods, and incentives. Aids for sustainable design must be easy and quick to use and understand. Some specific approaches to sustainable design include:

Mass balance analysis involves tracing the materials or energy in and out of an analysis area such as a manufacturing station, a plant or a watershed. Ideally, mass balances are based on measurements of inflows, inventories, and outflows (including products, wastes and emissions). Actually, all the data needed is rarely available or even consistent.

Green indices Green indices or ranking systems attempt to summarize various environmental impacts into a simple scale. The designer or decision-maker can then compare the green score of alternatives (materials, processes, etc.) and choose the one with the minimal environmental impact. This would contribute to products with reduced environmental impacts.

Design for disassembly and recycling(DFD/R) means making products that can be taken apart easily for subsequent recycling and parts reuse. Unfortunately, the economic costs associated with physically taking apart products to get at valuable components and materials often exceed the value of the materials. Reducing the time (and thus cost) of disassembly might reverse this balance. DFD/R software tools generally calculate potential disassembly pathways, point out the fastest pathway, and reveal obstacles to disassembly that can be "designed out".

Risk analysis is a means for tracing through the chances of different effects occurring. For example, the risk of toxic emissions is evaluated by estimating the amount and type of emissions, the transport in the environment, the ecological and human exposure, and the likely damage.

Material selection and label advisors Material selection guidelines attempt to guide designers towards the environmentally preferred material. Label advisors are generally marks on materials or products that reveal information about the material content relevant to materials handling and waste management. Embedding labels into a material is preferable to a separate label material attached with an adhesive.

Developing and marketing sustainable products is a concrete step towards sensible resource use and environmental protection and towards sustainable economic development. Sustainable products imply more efficient resource use, reduced emissions, and reduced waste, lowering the social cost of pollution control and environmental protection. Sustainable products promise greater profits to companies by reducing costs (reduced material requirements, reduced disposal fees, and reduced environmental cleanup fees), and raising revenues through greater sales and exports. Designing sustainable products offers much to the current generation, as well as providing future generations with a planet that will enable them to survive and prosper.

New Words and Expressions

1. sustainable product design
    可持续产品设计
2. cradle-to-grave   摇篮到坟墓
3. hazardous   冒险的,有危险的
4. benign   有益健康的,良性的
5. dump   倾倒,丢弃,倾卸
6. remediation   修复,补救,纠正
7. solvent substitution   溶剂取代
8. dissipation   消失,消散,消耗
9. life cycle assessment (LCA)
    生命周期评价
10. material flows and cycles   物流循环跟踪
11. flawed   有缺陷的
12. landfill   废弃物填埋场,填埋的废弃物
13. compile   收集,编辑,编制
14. closed-loop   闭环,闭合线路
15. green indices   绿色指数
16. cast aside   丢弃,废除

Notes

(1) Consumers cast aside products, usually with only minimal re-manufacturing or recycling. Recognition of theses problems inspired environmental engineering applications to clean up past pollution (called remediation) and ongoing wastes streams (called waste treatment).

消费者丢弃的废品一般只有少量被再制造或循环使用，认识到这些问题之后，环境工程技术的研究开始转向对已造成污染的治理（环境修复）和当前污染治理（废物处理）。

(2) Clean up are still needed in many cases. But design changes can often be more effective at reducing environmental burdens and more efficient at reducing costs than traditional "end-of-the-pipe" clean up strategies.

末端治理技术至今在很多场合仍然需要，但是如果改变产品的设计方法，对减少环境污染和降低成本更为有效。

(3) Changing design procedures is particularly difficult because designers face many conflicting objectives, uncertainties, and a work environment demanding speed and cost effectiveness.

产品设计方法的革新是一项艰巨任务，因为设计人员需要面对设计目标冲突、不确定性、设计时间和成本效益等诸多问题。

(4) Source reduction refers to any change in the design, manufacture, purchase, or use of materials or products (including packaging) to reduce their amount or toxicity before they become municipal solid waste.

资源减量化是指在产品的设计、制造、采购等过程中，减少原材料（包括包装材料）和有毒材料的用量，减少废物产生量。

(5) Mass balance analysis involves tracing the materials or energy in and out of an analysis area such as a manufacturing station, a plant or a watershed. Ideally, mass balances are based on measurements of inflows, inventories, and outflows (including products, wastes and emissions).

物料平衡分析是针对某一工厂、制造车间或生产单元，对输入和输出的物料和能源情况进行分析、核算的过程。从理论上讲，物料平衡的基础是对系统输入量、产品库存和输出量（包括产品、废物和耗散量）的测定。

(6) Reducing the time (and thus cost) of disassembly might reverse this balance.

单纯追求减少产品的拆卸时间（由此降低拆卸成本），往往会亏本。

❖ Writing Training

如何制作简历——简历的内容与制作时的注意事项

（一）简历内容

建议按照以下内容及顺序来写简历。

（1）个人基本信息，包括姓名及联系方式等（Basic personal information）；

(2) 求职意向(Job-seeking intention);
(3) 教育经历(Educational background);
(4) 工作及实习经历(Work and internship experiences);
(5) 项目经历(如果有的话)(Project experiences);
(6) 社会实践(如果有的话)(Social practice);
(7) 所获奖励或证书(Reward and Certificate);
(8) 外语及计算机技能等(English and computer skills)。

结合自己的背景和特点,对相关内容的顺序可以作相应的调整。调整原则:重要的、能突出自己的优势和职位要求的内容往前排,不太重要的内容往后排。

(二) 简历制作中应该注意的问题

在制作简历时,要时刻记住你是在一个商业环境中推销自己。如果你是个初出校园的毕业生,就应该通过简历让聘用方感到你对未来职业生涯已作好了充分准备。有针对性的内容和规范专业的格式都是你的简历必须具备的。

在形式上,一份专业的简历应当符合以下要求:
(1) 要疏密有致、主次分明;
(2) 控制在一页范围内;
(3) 各种级别的字体要选择适当;
(4) 不要使用框格;
(5) 不要出现"Resume"字样;
(6) 不要以学校的标识(Logo)和名称作为页眉。

打印简历时,应该注意以下几点:
(1) 要使用 80 g 左右的白色或奶白色纸张;
(2) 不要选择彩色打印;
(3) 不要选择喷墨打印;
(4) 不要用复印的简历。

❖ Reading Material

Cleaner Production Strategies

1. Introduction

Over the years, industrialized nations have progressively taken different approaches to dealing with environmental degradation and pollution problems, by:

- Ignoring the problem;
- Diluting or dispersing the pollution so that its effects are less harmful or apparent;
- Controlling pollution using "end-of-pipe" treatment;
- Preventing pollution and waste at the source through a "Cleaner Production" approach.

The gradual progression from "ignore" through to "prevent" has culminated in the realization that it is possible to achieve economic savings for industry as well as an improved environment for society. This, essentially, is the goal of Cleaner Production (see Figure 20.5).

The term Cleaner Production (CP) was defined by UNEP in 1998 as: "The continuous application of an integrated, preventive environmental strategy towards processes, products and services in order to increase overall efficiency and reduce risks for humans and the environment."

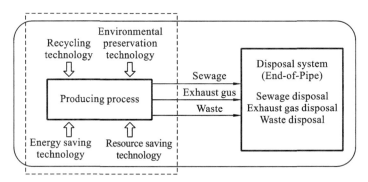

Figure 20.5 Framework of cleaner production

- For production processes: Cleaner production includes conserving raw materials and energy, eliminating toxic raw materials, and reducing the quantity and toxicity of all emissions and wastes;
- For products: Cleaner production includes the reduction of negative impacts along the life cycle of a product, from raw material extraction to its ultimate disposal;
- For services: Cleaner production is to incorporate environmental concerns into designing and delivering services.

Cleaner production is a key concept that results in environmentally sustainable and economically viable business practices. It includes not only the conventional technologies for each facility and measure (hard technology), but the technologies for operation and management methods (soft technology), based on the idea of reducing the environmental burden in every process from extracting of raw materials to disposal of products and reuse (see Figure 20.6).

According to a report *"Cleaner Production and Eco-efficiency"* published by UNEP/WBSCD (World Business Council for Sustainable Development) in 1998, governments need to carry out a review of legislation while stakeholders need to cooperate and coordinate in order to promote cleaner production, the result of which would be less resource usage and waste by a society as a whole and greater creation of value.

Currently, the collection and dissemination of the technology information are regarded as one of the important international subjects and countries are making efforts to tackling the issue. Cleaner production reduces consumption of raw materials and resources as well as production of solid waste. In most of the cases, cleaner production can minimize or even completely eliminate the need for investments of end of pipe solutions such as waste or water treatment.

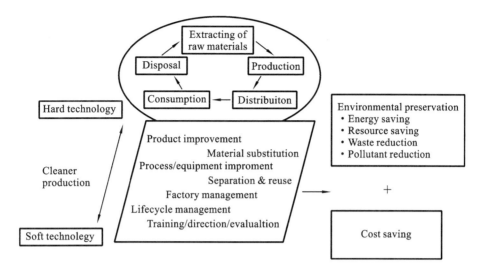

Figure 20.6 Cleaner production strategy

2. Cleaner production principles

Cleaner production integrates three underlying principles:

1) The preventive principle

It is cheaper and more effective to prevent environmental damage than to attempt to manage or "cure" it. Prevention involves using safer chemicals and eliminating hazardous chemicals, including though substitution, with effective non-hazardous alternatives. Where toxic chemicals are currently used, the elimination of spills, accidents and fugitive releases is required while safer alternatives are researched and implemented.

2) The public participation principle

Public access to information about emissions and releases of hazardous chemicals from manufacturing facilities, the amounts and types of chemicals and materials used in production processes and the chemical ingredients in products is necessary to move to safer alternatives and can hasten the adoption of clean production.

3) The holistic principle

Clean production is an integrated approach to production, constantly asking what happens throughout the life cycle of the chemical or product. It is necessary to think in terms of integrated systems, which is how the living world functions. Otherwise new problems may be created by trying to solve old ones, such as changing the manufacturing process to stop the direct discharge of hazardous chemicals in wastewaters by redirecting them to a waste water treatment plant which is unable to adequately treat many of the chemicals, wherein those chemicals are simply transferred to the sludge, thus generating a new hazardous waste stream.

3. Cleaner Production Assessment

To be able to identify cleaner production options, it is necessary to carry out a cleaner production assessment. The cleaner production assessment focuses on:

- Where waste and emissions are generated;

- Why waste and emissions are generated;
- How waste and emissions can be minimized in your company.

Cleaner production assessment is a useful tool to systematically investigate the existing production and to identify opportunities for improving the production or the products. Steps towards cleaner production are as following:

Step 1: Getting started

Ensure top management commitment; Form a cleaner production team; List process steps and identify waste streams; Prepare process flow charts; and Select focus areas.

Step 2: Analyzing process steps

Make material and energy balances; Characterize waste streams; Assign costs to waste streams; and Identify causes of waste generation.

Step 3: Generating cleaner production options

Generate workable cleaner production options; and Sort options into: "directly implementable"; "needs further study"; and "rejected options".

Step 4: Selecting cleaner production options

Analyze technical feasibility of cleaner production options; Analyze economic viability of cleaner production options; Analyze environmental feasibility of cleaner production options; and Select cleaner production options for implementation.

Step 5: Implementing CP options

Make a cleaner production action plan; Implement the cleaner production options.

Step 6: Maintaining cleaner production

Monitor and evaluate results; Report cleaner production results; Prepare for a new cleaner production assessment; and continuously integrate cleaner production activities into daily management.

It is important to stress that cleaner production is about attitudinal as well as technological change. In many cases, the most significant cleaner production benefits can be gained through lateral thinking, without adopting technological solutions. A change in attitude on the part of company directors, managers and employees is crucial to gaining the most from cleaner production.

New Words and Expressions

1. cleaner production 清洁生产
2. stakeholder 投资者,利益相关者
3. eco-efficiency 生态效益,生态效率
4. spill 溢出,泼出,泄漏
5. stakeholder 投资者,利益相关者
6. holistic 全部的,整体的
7. generating cleaner production options 产生清洁生产方案
8. ingredient 组成部分,配料
9. kiln 窑,窑炉

Lesson 21 Automobile Engineering

Text

Since the beginning of the twentieth century, automobiles have entered the lives and livelihoods of almost everyone. An automobile is a self-propelled vehicle driven by an internal combustion engine and is used for transportation of passengers and goods on ground. Different types of automobiles include cars, buses, trucks, vans, and motorcycles, with cars being the most popular. The term is derived from Greek "autos" (self) and Latin "movére" (move), referring to the fact that it "moves by itself". An automobile has seats for the driver and, almost without exception, one or more passengers. Today it is the main source of transportation across the world.

The major components which are used in an automobile include: the engine, chassis, body and the electrical system.

The engine is a power plant or a motor, which provides power to drive the automobile. In most automobile engines, the burning of fuel develops the pressures. These pressures are transmitted to the crankshaft by the pistons and connecting rods and torque is produced which sets the crankshaft in motion. The torque produced by the engine is transmitted through the drive line to the road wheels to propel the vehicle. A number of systems are necessary to make an engine work. A lubrication system is needed to reduce friction and prevent engine wear. A cooling system is required to keep the engine's temperature within safe limits. The engine must be provided with the correct amount of air and fuel by a fuel system. The mixture of air and fuel must be ignited inside the cylinder at just the right time by an ignition system. Finally, an electrical system is required to operate the starting motor that starts the engine and to provide electrical energy to power engine accessories.

The chassis is an assembly of the frame, suspension, power train, steering, and braking system. The automobile frame is usually made up of a number of square or box-shaped steel members welded or riveted together to give the final shape which is strong enough to support the weight of the body and other components. The function of the suspension system is to absorb vibrations due to the up and down motion of wheels, caused by the irregularities in the road surface. The springs, connecting linkages, and shock absorber comprise the suspension system of a vehicle. Absorption and damping of vibrations protects passengers from discomfort caused by shocks. The power train carries the power that the engine produces to the car wheels. It consists of the clutch (on cars with a manual transmission), transmission (a system of gears that increases the turning effort of the engine to move the automobile), drive shaft, differential and rear axle. The steering system is used for changing the direction of the vehicle. The major requirements in any steering mechanism are that it should be precise and easy to handle, and that the front wheels should have a tendency to return to the straight-ahead position after a turn. A gear mechanism, which is known as steering gear, is used in this system to increase the steering effort provided by the driver. This system makes the vehicle steering very easy as the driver does not have to put in much effort.

Brakes are required for slowing down or stopping a moving vehicle. The braking system is essential for the safety of passengers, and passers-by on roads. The braking system may be operated mechanically or hydraulically. Most of the braking systems in use today are of the hydraulic type. Brakes are hydraulic so that failures are slow leaks, rather than abrupt cable breaks. All brakes consist of two members, one rotation and the other stationary. There are various means by which the two members can be brought in contact, thus reducing the speed of the vehicle. All vehicles must be fitted with at least two independent systems. They were once called the service brake and the emergency brake. Now they are usually referred to as the foot brake and the park brake. Most light vehicles use a foot brake that operates through a hydraulic system on all wheels and a hand operated brake that acts mechanically on the rear wheels only. One common use of the hand brake system is to hold the vehicle when it is parked. The systems are designed to be independent so that if one fails, the other is still available.

The main purpose of the body work is to provide accommodation for the driver and passengers, with suitable protection against wind and weather. The degree of comfort provided depends upon the type of car and its cost. The body on the first automobiles was little more than a platform with seats attached. It gradually developed into a closed compartment complete with roof and windows. The modern automobile body is constructed of sheet steel formed to the required shape in giant punch presses. Most of the body components are welded together to form a light rattle-free unit. All steel surfaces must be treated and painted. Primarily this is to give protection against rust corrosion, and secondly to improve the appearance. External trim, made of stainless steel, chromium-plated brass or plastics, embellishes the body and appeals to the eye by providing a contrast with the plain coloured surface. Automobiles may also be classed on the basis of their body style(Figure 21.1).

The electrical system provides energy to operate a starting motor and to power all the accessories. The main components of the electrical system are a battery, an alternator, a starting motor, ignition coil and heater. This system starts the engine when the ignition switch is turned on. It makes the spark that ignites the compressed air-fuel mixture. It also operates the head lights, indicator lights, brake lights parking lights, wipers, and also air conditioning, radio and cassette recorders if fitted in cars. The battery provides electrical energy for starting, then once the engine is running, the alternator supplies all the electrical components of the vehicle. It also charges the battery to replace the energy used to start the engine.

Lesson 21 Automobile Engineering

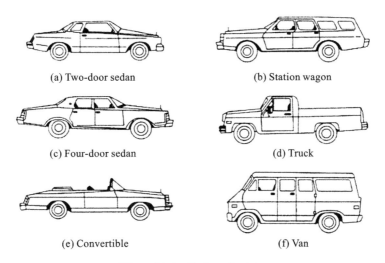

Figure 21.1 Bodywork styles

New Words and Expressions

1. automobile   汽车
2. livelihood   生计,谋生
3. propel   推进,驱使
4. crankshaft   曲轴
5. piston   活塞
6. connecting rod   连杆
7. lubrication   润滑
8. cylinder   汽缸
9. chassis   底盘
10. vibration   振动,摆动
11. spring   弹簧
12. linkage   联动装置,连接,连杆组
13. shock absorber   减振器、缓冲器
14. damp   阻尼,使衰减,控制
15. clutch   离合器
16. transmission   变速器,传输,传动装置
17. differential   差动(齿轮),差速器
18. punch press   冲床,冲压机
19. chromium-plated brass   镀铬黄铜
20. embellish   装饰,修饰
21. accessory   附件,配件,辅助设备
22. wiper   雨刷
23. sedan   私家轿车
24. station wagon   旅行车
25. van   有篷货车
26. convertible   敞篷车

Notes

The chassis is an assembly of the frame, suspension, power train, steering, and braking system. The automobile frame is usually made up of a number of square or box-shaped steel members welded or riveted together to give the final shape which is strong enough to support the weight of the body and other components.

底盘由车架、悬挂系统、传动系统、转向系统和制动系统组成。车架通常由若干方钢或箱形型钢构件焊接或铆接在一起形成,最终的轮廓有足够的强度来支撑车身和其他部件的重量。

❖ Writing Training

如何制作简历——写作简历的要点

写作简历时应注意以下几点。

(1) 简历定位明确。雇主们花时间和精力来到招聘现场阅读简历、筛选人才,目的就是想知道你可以为他们做什么,而不是来欣赏你的简历的文采和笔锋,所以含糊的、笼统的、毫无针对性的简历会使雇主感到茫然,也会使你失去很多机会。应明确你到底能干什么、最能干的是什么。如果你有多个目标,最好写上多份不同的简历,在每一份上突出重点,这将使你的简历更有机会脱颖而出。

(2) 使用"从事事件:结果"这种格式。仅有漂亮的外表而无内容的简历是不会吸引人的注意力的,内容决定一切。所以简历中一定要有过硬的内容,特别要突出你的能力、成就以及取得的经验,这样才会使你的简历富有特色而更加出众。仔细分析你的能力并阐明你能够胜任这份工作,强调以前完成的事件然后要写上结果。当然对完成什么事件是应该有所选择的,突出用人单位会欣赏的经验和能力,不要不着边际地写上一大堆。

(3) 让简历醒目。简历的外表不一定要很华丽,但它至少要清楚、醒目。审视一下简历的空白处,用这些空白处和边框来强调你的正文,或使用各种字体格式,如斜体、大写、下划线、首字突出、首行缩进或加符号等办法。要用电脑来打印你的简历。

(4) 尽量使你的简历简短,有可能只使用一张纸。雇主一般只会花 30 秒来扫视一下你的简历,然后决定是否要面试你,所以简历越简练、精悍,效果越好。如果你有很长的职业经历,一张纸写不下,试着组织出一张最有说服力的简历,删除那些无用的东西。

(5) 力求精确。阐述你的技巧、能力、经验时要尽可能准确,不夸大也不误导,不要模糊处理,同时要确信你所写的内容与你的实际能力及工作水平相符。

(6) 强调成功经验。雇主们想要能证明你实力的证据,要能证明你以前的成就并说明获得这些成就的原因和经验,一定要客观和准确地说明你在取得这些成就的过程中有什么创新、有什么特别的办法,这样的人才一般会受到用人单位的青睐。

(7) 使用有影响力的词汇。使用诸如:证明的、分析的、有创造力的和有组织的等这样的词汇,这样可以提高简历的说服力。尽量每句都用到这种词汇。

(8) 不要写上个人爱好。如果招聘单位没有特别的要求,不要把个人爱好写在简历上,因为现在许多用人单位对纯个人信息没有要求,在简历上写上个人信息如婚姻状况、血型等已不再是必须,许多公司都乐意接受没有纯个人信息的简历。有些学生为了体现自己的高雅爱好,往往添上一笔,而用人单位一般都认为,许多个人爱好仅仅停留在爱好的层面而已,对工作并没有什么大用。当然,如果应聘的职位和你的爱好关系比较紧密的话也不妨写上,比如应聘记者时,不妨写上爱好读书、写作等等,也没有什么坏处。

(9) 检查与完善。写完以后,再检查一下:你的简历是否清楚并能够让雇主尽快知道你的能力?是否写清了你的能力?是否写清了你要求这份工作的基础?有东西可删除吗?尽力完善你的简历,直到最好为止。

❖ Reading Material

Internal Combustion Engine

An engine in an automobile is the source of power. Any engine which derives heat energy from the combustion of fuel and converts this energy into mechanical work is termed as a heat engine. If combustion occurs within the cylinder, the engine is called an internal combustion engine. If combustion takes place outside the cylinder, the engine is called an external combustion engine. Engines used in automobiles are internal combustion heat engines. When the fuel burns in the cylinder of an engine, the air inside can be raised to a very high temperature and so will reach a correspondingly high pressure. These pressure drives the pistons. The pistons turn a crankshaft to which they are attached. The rotation force of the crankshaft makes the automobile's wheels turn through drive line.

The piston fits closely inside the cylinder. Ideally it would be perfectly gas-tight yet perfectly free to move up and down inside the cylinder. The connection rod connects the piston to the crankshaft. At the piston end is a pin called the gudgeon pin which is fitted into holes in the piston and connection rod to couple them together. The crankshaft is the main shaft of the engine and is carried in bearings in the crankcase. Offset from the main part of the shaft is the crank pin on which the connecting rod is fitted and is free to turn.

The crankshaft can be rotated by pushing the piston up and down in the cylinder. Starting with position shown in Figure 21.2, the crankshaft rotates clockwise as the piston is pushed downwards until the piston reaches the lowest point of its travel. At this point the crank pin will be directly under the centre of the crankshaft, and the centres of the gudgeon pin, crank pin and crankshaft will all lie in a straight line. In this position pressure on the piston will have no turning effect on the crankshaft, and this position is therefore called a dead centre. Another dead centre occurs when the piston is at the extreme top of its travel. These two dead centres, which are known as bottom dead centre (BDC) and top dead center

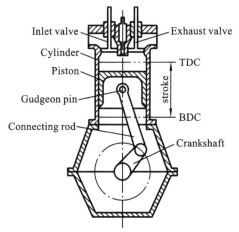

Figure 21.2 **The main parts of an engine**

(TDC) respectively, mark the extreme limits of the piston's travel. The volume between BDC and TDC is called swept volumes. The volume of the space above the piston when it is at TDC is called clearance volume. Engine capacity is the swept volume of all the cylinders e. g. a four-cylinder engine having a capacity of two liters(2000 cm^3) has a cylinder swept volume of 500 cm^3. Movement of the piston from one dead centre to another is called a stroke, and there are tow strokes of the piston to every revolution of the crankshaft.

In an engine the air can be heated to a very high temperature and a correspondingly high pressure created inside the cylinder, thus exerting a considerable force on the (its upper surface) piston. Pressure above the piston can only push it downwards. The piston is returned up the cylinder by the rotation of a wheel with a heavy rim—called a flywheel—fitted to the crankshaft. Once this flywheel has been made to turn, it will continue to rotate for several revolutions.

Internal combustion engines can use any one of a variety of fuels. Petrol is a liquid refined from crude petroleum, and is particularly suitable as a fuel for motor vehicles. Before petrol can be burnt it must be vaporized and mixed with a suitable quantity of air. This mixture must then be introduced into the cylinder. The control of gas movement is the duty of the valves: an inlet valve allows the new mixture to enter at the right time and an exhaust valve lets out the burnt gas after the gas has done its job.

The running of the engine involves the continuous repetition of four operations which make up what is called the cycle of operations. These operations are continuously repeated as long as the engine is running in the following order:(see Figure 21.3)

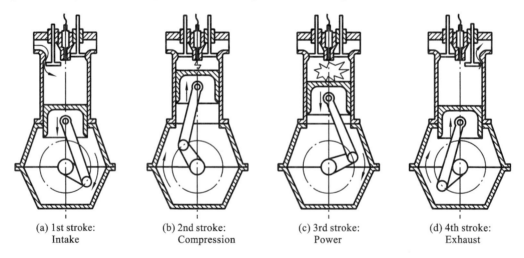

(a) 1st stroke: Intake   (b) 2nd stroke: Compression   (c) 3rd stroke: Power   (d) 4th stroke: Exhaust

Figure 21.3 The operation of a four-stroke cycle engine

1. Intake Stroke

The downward-moving piston increases the volume in the cylinder and draws in fresh air-fuel mixture though the open inlet valve.

2. Compression stroke

The upward-moving piston reduces the volume in the cylinder and compresses the air-fuel mixture. Shortly before TDC is reached, the spark plug ignites the compressed air-fuel mixture and thus initiates the combustion process. A higher compression ratio means better utilization of the fuel.

$$\text{Compression ratio} = \frac{\text{swept vol} + \text{clearance vol}}{\text{clearance vol}}$$

3. Power stroke

After the ignition spark at spark plug has ignited the compressed air-fuel mixture, the temperature increases as the result of combustion of the mixture. The pressure in the cylinder increases and forces the piston downwards. The piston transfers power to the crankshaft via the connecting rod.

4. Exhaust stroke

The upward-moving piston expels the combusted gases (exhaust gas) through the open exhaust valve. After this 4th stroke, the cycle is repeated.

As described above, an internal combustion engine in which the piston completes four separate strokes—intake, compression, power, and exhaust—during two separate revolutions of the engine's crankshaft, and one single thermodynamic cycle is a four-stroke engine.

New Words and Expressions

1. internal combustion engine 内燃机
2. rim 边,轮缘
3. drive line 传动(轴)系
4. flywheel 飞轮
5. gudgeon pin 活塞销
6. petrol 汽油
7. crank case 曲轴箱
8. crude petroleum 原油
9. crank pin 曲轴销,曲柄销
10. inlet valve 进气阀
11. bottom dead centre(BDC) 下止点
12. exhaust valve 排气阀
13. top dead centre(TDC) 上止点
14. intake stroke 进气行程
15. swept volume 有效容积
16. compression stroke 压缩行程
17. clearance volume 余隙容积,燃烧室容积
18. power stroke 做功行程
19. engine capacity 发动机排量
20. exhaust stroke 排气行程
21. liter 升
22. spark plug 火花塞
23. stroke 冲程,行程
24. compression ratio 压缩比
25. revolution 旋转

Lesson 22 Engineering Equipment

Text

A machine tool is a machine for shaping or machining metal or other rigid materials, usually by cutting, boring, grinding, shearing or other forms of deformation. Machine tools employ some sort of tool that does the cutting or shaping. All machine tools have some means of constraining the workpiece and provide a guided movement of the parts of the machine. Thus the relative movement between the workpiece and the cutting tool (which is called the toolpath) is controlled or constrained by the machine to at least some extent, rather than being entirely "offhand" or "freehand".

The precise definition of the term machine tool varies among users. It is safe to say that all machine tools are "machines that help people to make things", although not all factory machines are machine tools.

Today machine tools are typically powered other than by human muscle (e. g., electrically, hydraulically, or via line shaft), used to make manufactured parts (components) in various ways that include cutting or certain other kinds of deformation.

Broaching is a machining process that uses a toothed tool, called a broach, to remove material. There are two main types of broaching: linear and rotary. In linear broaching, which is the more common process, the broach is run linearly against a surface of the workpiece to effect the cut. Linear broaches are used in a broaching machine, which is also sometimes shortened to broach. In rotary broaching, the broach is rotated and pressed into the workpiece to cut an axis symmetric shape. A rotary broach is used in a lathe or screw machine. In both processes the cut is performed in one pass of the broach, which makes it very efficient.

Broaching is used when precision machining is required, especially for odd shapes. Commonly machined surfaces include circular and non-circular holes, splines, keyways, and flat surfaces. Typical workpieces include small to medium sized castings, forgings, screw machine parts, and stampings. Even though broaches can be expensive, broaching is usually favored over other processes when used for high-quantity production runs.

Broaches are shaped similar to a saw, except the teeth height increases over the length of the tool. Moreover, the broach contains three distinct sections: one for roughing, another for semi-finishing, and the final one for finishing. Broaching is an unusual machining process because it has the feed built into the tool. The profile of the machined surface is always the inverse of the profile of the broach. The rise per tooth (RPT), also known as the step or feed per tooth, determines the amount of material removed and the size of the chip. The broach can be moved relative to the workpiece or vice-versa. Because all of the features are built into the broach no complex motion or skilled labor is required to use it. A broach is effectively a collection of single-point cutting tools arrayed in sequence, cutting one after the other; its cut is analogous to multiple passes of a shaper.

A drill press (also known as horizontal drill, vertical drill, or bench drill) is a fixed style of drill that may be mounted on a stand or bolted to the floor or workbench. Portable models with a magnetic base grip the steel workpieces they drill. A drill press consists of a base, column (or pillar), table, spindle (or quill), and drill head, usually driven by an induction motor. The head has a set of handles (usually 3) radiating from a central hub that, when turned, move the spindle and chuck vertically, parallel to the axis of the column. The table can be adjusted vertically and is generally moved by a rack and pinion; however, some older models rely on the operator to lift and reclamp the table in position. The table may also be offset from the spindle's axis and in some cases rotated to a position perpendicular to the column. The size of a drill press is typically measured in terms of swing. Swing is defined as twice the throat distance, which is the distance from the center of the spindle to the closest edge of the pillar. For example, a 16 in(about 410 mm) drill press has an 8 in(about 200 mm) throat distance.

A gear shaper is a machine tool for cutting the teeth of internal or external gears. The name shaper relates to the fact that the cutter engages the part on the forward stroke and pulls away from the part on the return stroke, just like the clapper box on a planer shaper.

The cutting tool is also gear shaped having the same pitch as the gear to be cut. However, number of cutting teeth must be less than that of the gear to be cut for internal gears. For external gears the number of teeth on the cutter is limited only by the size of the shaping machine. For larger gears the blank is usually gashed to the rough shape to make shaping easier.

The principal motions involved in rotary gear shaper cutting are of the following:

• Cutting motion: The downward linear motion of the cutter spindle together with the cutter.

• Return stroke: The upward linear travel of the spindle and cutter to withdraw the latter to its starting position.

• Indexing motion: Slow speed continuous rotation of the cutter spindle and work spindle to provide circular feed, the two speeds being regulated through the change gears such that against each rotation of the cutter and the gear blank revolves through n/N revolution, where n—number of teeth of the cutter.

Hobbing is a machining process for making gears, splines, and sprockets on a hobbing machine, which is a special type of milling machine. The teeth or splines are progressively cut into the workpiece by a series of cuts made by a cutting tool called a hob. Compared to other gear forming processes it is relatively inexpensive but still quite accurate, thus it is used for a broad range of parts and quantities.

It is the most widely used gear cutting process for creating spur and helical gears and more gears are cut by hobbing than any other process since it is relatively quick and inexpensive.

Honing is an abrasive machining process that produces a precision surface on a metal workpiece by scrubbing an abrasive stone against it along a controlled path. Honing is primarily used to improve the geometric form of a surface, but may also improve the surface texture.

Typical applications are the finishing of cylinders for internal combustion engines, air bearing spindles and gears. Types of hone are many and various but all consist of one or more abrasive stones that are held under pressure against the surface they are working on.

In everyday use, a honing steel is used to hone knives, especially kitchen knives, and is a fine process, there contrasted with more abrasive sharpening.

A lathe is a machine tool which rotates the workpiece on its axis to perform various operations such as cutting, sanding, knurling, drilling, or deformation with tools that are applied to the workpiece to create an object which has symmetry about an axis of rotation. The essential components of a lathe are depicted in the schematic diagram of Figure 22.1.

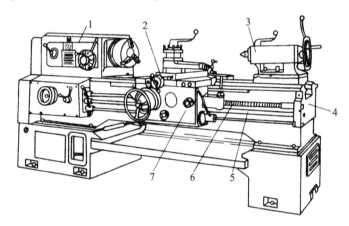

Figure 22.1 Schematic diagram of basic components of a lathe
1—headstcok; 2—toolpost; 3—tailstock; 4—bed; 5—feed shaft; 6—leadscrew; 7—carriage

Lathes are used in woodturning, metalworking, metal spinning, and glass-working. Lathes can be used to shape pottery, the best-known design being the potter's wheel. Most suitably equipped metalworking lathes can also be used to produce most solids of revolution, plane surfaces and screw threads or helices. Ornamental lathes can produce three-dimensional solids of incredible complexity. The material can be held in place by either one or two centers, at least one of which can be moved horizontally to accommodate varying material lengths. Other work-holding methods include clamping the work piece about the axis of rotation using a chuck or collet, or to a faceplate, using clamps or dogs.

Examples of objects that can be produced on a lathe include candlestick holders, gun barrels, cue sticks, table legs, bowls, baseball bats, musical instruments (especially woodwind instruments), crankshafts and camshafts.

A milling machine is a machine tool used to machine solid materials. Milling machines are often classed in two basic forms, horizontal and vertical, which refers to the orientation of the main spindle. Both types range in size from small, bench-mounted devices to room-sized machines. Unlike a drill press, which holds the workpiece stationary as the drill moves axially to penetrate the material, milling machines also move the workpiece radially against the rotating milling cutter, which cuts on its sides as well as its tip. Workpiece and cutter movement are precisely controlled to less than 0.001 in (0.025 mm), usually by means of precision ground slides and leadscrews or analogous technology. Milling machines may be manually operated, mechanically automated, or digitally automated via computer numerical control. Milling machines can perform a vast number of operations, from simple (e.g., slot and keyway cutting, planing, drilling) to complex (e.g., contouring, diesinking). Cutting fluid is often pumped to the cutting site to cool and lubricate the cut and to wash away the resulting swarf.

A shaper is a type of machine tool that uses linear relative motion between the workpiece and a single-point cutting tool to machine a linear toolpath. Its cut is analogous to that of a lathe, except that it is (archetypally) linear instead of helical. (Adding axes of motion can yield helical toolpaths, as also done in helical planing.) A shaper is analogous to a planer, but smaller, and with the cutter riding a ram that moves above a stationary workpiece, rather than the entire workpiece moving beneath the cutter. The ram is moved back and forth typically by a crank inside the column; hydraulically actuated shapers also exist.

Shapers are mainly classified as draw-cut, horizontal, universal, vertical, geared, crank, hydraulic, contour and traveling head. The horizontal arrangement is the most common. Vertical shapers are generally fitted with a rotary table to enable curved surfaces to be machined (same idea as in helical planing). The vertical shaper is essentially the same thing as a slotter (slotting machine), although technically a distinction can be made if one defines a true vertical shaper as a machine whose slide can be moved from the vertical. A slotter is fixed in the vertical plane.

Small shapers have been successfully made to operate by hand power. As size increases, the mass of the machine and its power requirements decrease, and it becomes necessary to use a motor or other supply of mechanical power. This motor drives a mechanical arrangement (using a pinion gear, bull gear, and crank, or a chain over sprockets) or a hydraulic motor that supplies the necessary movement via hydraulic cylinders.

A planer is a type of metalworking machine tool that uses linear relative motion between the workpiece and a single-point cutting tool to machine a linear toolpath. Its cut is analogous to that of a lathe, except that it is (archetypally) linear instead of helical. (Adding axes of motion can yield helical toolpaths.) A planer is analogous to a shaper, but larger, and with the entire workpiece moving on a table beneath the cutter, instead of the cutter riding a ram that moves above a stationary workpiece. The table is moved back and forth on the bed beneath the cutting head either by mechanical means, such as a rack and pinion drive or a leadscrew, or by a hydraulic cylinder.

A grinding machine, often shortened to grinder, is a machine tool used for grinding, which is a type of machining using an abrasive wheel as the cutting tool. Each grain of abrasive on the wheel's surface cuts a small chip from the workpiece via shear deformation.

Grinding is used to finish workpieces which must show high surface quality (e. g. , low surface roughness) and high accuracy of shape and dimension. As the accuracy in dimensions in grinding is on the order of 0. 000025 mm, in most applications it tends to be a finishing operation and removes comparatively little metal, about 0. 25 to 0. 50 mm depth. However, there are some roughing applications in which grinding removals high volumes of metal quite rapidly. Thus grinding is a diverse field.

New Words and Expressions

1. machine tool　机床
2. workpiece　工件
3. cutting tool　切削刀具
4. broaching　拉削
5. drill press　钻床
6. gear shaper
7. cutting motion　切削运动
8. return stroke　回程
9. indexing motion　转位运动
10. hobbing machine　滚齿机
11. honing　珩磨
12. lathe　车床
13. milling machine　铣床
14. shaper　牛头刨床
15. planer　龙门刨床
16. grinding machine　磨床

Notes

（1）A machine tool is a machine for shaping or machining metal or other rigid materials, usually by cutting, boring, grinding, shearing or other forms of deformation.

机床是一种用来加工成形金属或者其他硬质材料的机器,通常用来进行车削、钻孔、磨削、切断或其他的成形加工。

（2）Indexing motion：Slow speed continuous rotation of the cutter spindle and work spindle to provide circular feed, the two speeds being regulated through the change gears such that against each rotation of the cutter and the gear blank revolves through n/N revolution, where n—number of teeth of the cutter.

转位运动:刀盘主轴的工作主轴慢速旋转来进行循环进给,旋转速度通过变速齿轮来控制。被加工齿坯与刀具旋转方向相反,其速度根据 n/N 的大小来设定,其中 n 为插齿刀的齿数。

（3）The teeth or splines are progressively cut into the workpiece by a series of cuts made by a cutting tool called a hob. Compared to other gear forming processes it is relatively inexpensive but still quite accurate, thus it is used for a broad range of parts and quantities.

通过滚刀,一系列的齿或槽在工件上被加工出来。和其他的齿轮成形工艺相比,滚齿的费用相对是比较便宜的,而且精度也很好,所以,它可用于零件的大批量加工。

❖ Writing Training

如何制作简历——简历中常用的词汇和简历模板

- 简历中常用的词汇和简历模板

1. 教育(Education)类

学历 educational history	教授 professor
教育程度 educational background	副教授 associate professor
课程 curriculum	讲师 lecturer
主修 major	助教 teaching assistant
辅修 minor	研究员 research fellow
专门课程 specialized courses	助理研究员 research assistant
所学课程 courses taken	论文导师 supervisor
所学课程 courses completed	中学校长 principal(美)
特别训练 special training	中小学校长 headmaster(英)
社会实践 social practice	小学校长 master(美)
业余工作 part-time jobs	教务长 dean of students
暑期工作 summer jobs	教导主任 dean of students
假期工作 vacation jobs	教师 teacher
进修课程 refresher course	及格 pass
体育活动 extracurricular activities	不及格 fail
娱乐活动 recreational activities	分数 marks
学术活动 academic activities	分数 grades
社会活动 social activities	分数 scores
奖励 rewards	考试 examination
奖学金 scholarship	班长 monitor
"三好"学生 "Three Goods" student	副班长 vice-monitor
优秀团员 excellent League member	学习委员 commissary in charge of studies
优秀干部 excellent leader	文娱委员
学生会 student council	commissary in charge of entertainment
脱产培训 off-job training	体育委员
在职培训 in-job training	commissary in charge of sports
学制 educational system	劳动委员 commissary in charge of physical labor
学年 academic year	党支部书记 Party branch secretary
学期 semester(美)	团支部书记 League branch secretary
学期 term(英)	组织委员 commissary in charge of organization
校长 president	宣传委员 commissary in charge of publicity
副校长 vice-president	大学肄业生;(尚未取得学位的)大学生 undergraduate
教务员 academic dean	
系主任 department chairman	大学四年级学生;高中三年级学生 senior

大学三年级学生;高中二年级学生 junior
大学二年级学生;高中一年级学生 sophomore
大学一年级学生 freshman
实习生 intern
奖学金生 prize fellow

2. 常见职位名称(The titles of common positions)

行政助理 administration assistant
行政主管 administrator
学徒 apprentice
副经理 assistant manager
副厂长 assistant production manager
业务经理 business manager
总工程师 chief engineer
文员(文书) clerk
董事 director
电气工程师 electrical engineer
行政董事 executive director
行政秘书 executive secretary
领班,组长 foreman
总经理 general manager
低级文员(低级职员) junior clerk
经理 manager
市场部主任 marketing executive
市场部经理 marketing manager
市场部办公室主任 marketing officer
机械工程师 mechanical engineer
写字楼助理(办事员) office assistant
厂长 plant manager, production manager
品质控制员(质量检查员) quality controller
接线生(接线员) receptionist
销售工程师 sales engineer
销售主任 sales executive
销售经理 sales manager
营业代表 sales representative
推销员 salesman
秘书 secretary
高级文员,高级职员 senior clerk
熟练技工 skilled worker
副经理 sub-manager
主管 supervisor
测量员 surveyor
技术员 technician
翻译员 translator
打字员 typist

3. 证书(Certificates)

英语四级证书
College English Test 4 (CET4)
英语六级证书
College English Test 6 (CET6)
制图员证 cartographer
电工 electrician
钳工 fitter
电焊工 welder

- **求职简历模板**

XXXXXXX

Mobile Phone: ######   E-mail: ######@###.com

Home address: ******

Objective

Mechanical Engineer

Education Background

2005.9—2009.6   **Qingdao Technical College   Qingdao, China**
Bachelor of Mechanical Engineering, expected 2009

Major Courses: Graphing of Engineering, Principle of Mechanics, Mechanics of Materials, etc

Professional Experience

2008.1—2008.2	**Bank of China,Shanghai ZhengDa Branch [Internship]**
	• Good working knowledge of banking practice, communicating with customers
2007.7—2007.8	**United Securities Co.,Ltd,Shanghai West ChangJiang Road Sales Department [Associate Investment Manager]**
	• Learned skills in market analyzing and professional sales techniques
	• Successfully developed some potential customers for the company
2006.9—2007.4	**Daylight (China) Co.,Ltd [Sales Representative]**
	• Established a work team and led discussion in meetings when needed
	• Mastered knowledge of the products and finally rank 2nd among the entire community
2006.7~2006.8	**Shanghai Horizon Market Research Co.,Ltd [Researcher]**
	• Successfully arranged and conducted phone and indoor interviews across the city
	• Worked with different teams and finished questionnaires seriously

Social Experience

• Assisted my mentor to supervise the research work of two senior students(2005)

• Harmonized working between the cooperative research institute and our lab

• (undergraduate) vice class president 3 years

Honors and Scholarships

- Academic: National Scholarship, 1st Scholarship (once, 1 student/semester) in 2007—2008

  Scholarship for Outstanding Students,3rd Scholarship(twice,top 10%) in 2006—2007

  Excellent individual of Summer Practice and 2nd Place for Social Articles in 2006—2007

- Social: Excellent Marketing Manager in Human Resources Services Company of SIFT

  Excellent individual of Honor of Work for Study in 2006—2007

  Excellent Representative in Sports,Law School in 2007—2008

Certificates and Skills

- Language: The Intermediate Certification of Oral (verbal) Interpretation of Shanghai CET-6:525

- Computer: Certification of middle-level skills of computer operation of Shanghai

❖ Reading Material

Cutting Tools

In the context of machining, a cutting tool (or cutter) is any tool that is used to remove material from the workpiece by means of shear deformation. Cutting may be accomplished by single-point or multipoint tools. Single-point tools are used in turning, shaping, planing and similar operations, and remove material by means of one cutting edge. Milling and drilling tools are often multipoint tools. Grinding tools are also multipoint tools. Each grain of abrasive functions as a microscopic single-point cutting edge (although of high negative rake angle), and shears a tiny chip.

Cutting tools must be made of a material harder than the material which is to be cut, and the tool must be able to withstand the heat generated in the metal-cutting process. Also, the tool must have a specific geometry, with clearance angles designed so that the cutting edge can contact the workpiece without the rest of the tool dragging on the workpiece surface. The angle of the cutting face is also important, as is the flute width, number of flutes or teeth, and margin size. In order to have a long working life, all of the above must be optimized, plus the speeds and feeds at which the tool is run.

To produce quality perfect, a cutting tool must have three characteristics:
- Hardness — hardness and strength at high temperatures.
- Toughness — toughness, so that tools don't chip or fracture.
- Wear resistance — having acceptable tool life before needing to be replaced.

Cutting tool materials can be divided into two main categories: stable and unstable.

Unstable materials (usually steels) are substances that start at a relatively low hardness point and are then heat treated to promote the growth of hard particles (usually carbides) inside the original matrix, which increases the overall hardness of the material at the expense of some its original toughness. Since heat is the mechanism to alter the structure of the substance and at the same time the cutting action produces a lot of heat, such substances are inherently unstable under machining conditions.

Stable materials (usually tungsten carbide) are substances that remain relatively stable under the heat produced by most machining conditions, as they don't attain their hardness through heat. They wear down due to abrasion, but generally don't change their properties much during use.

Most stable materials are hard enough to break before flexing, which makes them very fragile. To avoid chipping at the cutting edge, most tools made of such materials are finished with a sightly blunt edge, which results in higher cutting forces due to an increased shear area. Fragility combined with high cutting forces results in most stable materials being unsuitable for use in anything but large, heavy and stiff machinery.

Unstable materials, being generally softer and thus tougher, generally can stand a bit of flexing without breaking, which makes them much more suitable for unfavorable machining conditions, such as those encountered in hand tools and light machinery. Some of the tool materials and their properties are listed in Table 22.1.

Table 22.1 The tool materials and their properties

Tool materials	Properties
Carbon tool steels	Unstable. Very inexpensive. Extremely sensitive to heat. Mostly obsolete in today's commercial machining, although it is still commonly found in non-intensive applications such as hobbyist or MRO machining, where economy-grade drill bits, taps and dies, hacksaw blades, and reamers are still usually made of it (because of its affordability). Hardness up to about 65 HRC. Sharp cutting edges possible
High speed steel (HSS)	Unstable. Inexpensive. Retains hardness at moderate temperatures. The most common cutting tool material used today. Used extensively on drill bits and taps. Hardness up to about 67 HRC. Sharp cutting edges possible
HSS cobalt	Unstable. Moderately expensive. The high cobalt versions of high speed steel are very resistant to heat and thus excellent for machining abrasive and/or work hardening materials such as titanium and stainless steel. Used extensively on milling cutters and drill bits. Hardness up to about 70 HRC. Sharp cutting edges possible
Cast cobalt alloys	Stable. Expensive. Somewhat fragile. Despite its stability it doesn't allow for high machining speed due to low hardness. Not used much. Hardness up to about 65 HRC. Sharp cutting edges possible
Cemented carbide	Stable. Moderately expensive. The most common material used in the industry today. It is offered in several "grades" containing different proportions of tungsten carbide and binder (usually cobalt). High resistance to abrasion. High solubility in iron requires the additions of tantalum carbide and niobium carbide for steel usage. Its main use is in turning tool bits although it is very common in milling cutters and saw blades. Hardness up to about 90 HRC. Sharp edges generally not recommended
Ceramics	Stable. Moderately inexpensive. Chemically inert and extremely resistant to heat, ceramics are usually desirable in high speed applications, the only drawback being their high fragility. Ceramics are considered unpredictable under unfavorable conditions. The most common ceramic materials are based on alumina (aluminium oxide), silicon nitride and silicon carbide. Used almost exclusively on turning tool bits. Hardness up to about 93 HRC. Sharp cutting edges and positive rake angles are to be avoided
Cermets	Stable. Moderately expensive. Another cemented material based on titanium carbide (TiC). Binder is usually nickel. It provides higher abrasion resistance compared to tungsten carbide at the expense of some toughness. Extremely high resistance to abrasion. Used primarily on turning tool bits although research is being carried on producing other cutting tools. Hardness up to about 93 HRC. Sharp edges generally not recommended

continue

Tool materials	Properties
Cubic boron nitride (CBN)	Stable. Expensive. Being the second hardest substance known, it is also the second most fragile. It offers extremely high resistance to abrasion at the expense of much toughness. It is generally used in a machining process called "hard machining", which involves running the tool or the part fast enough to melt it before it touches the edge, softening it considerably. Used almost exclusively on turning tool bits. Hardness higher than 95 HRC. Sharp edges generally not recommended
Diamond	Stable. Very Expensive. The hardest substance known to date. Superior resistance to abrasion but also high chemical affinity to iron which results in being unsuitable for steel machining. It is used where abrasive materials would wear anything else. Extremely fragile. Used almost exclusively on turning tool bits although it can be used as a coating on many kinds of tools. Sharp edges generally not recommended

New Words and Expressions

1. shear deformation 剪切变形
2. tiny chip 微小碎片
3. fragile 易碎的
4. binder 黏结剂

Lesson 23 Energy Saving Equipment

Text

There has been an enormous increase in the global demand for energy in recent years as a result of industrial development and population growth. Today, it is nearly twice what it was 30 years ago. By 2030, it may have risen by over 50 percent again, according to estimates by the International Energy Agency (IEA). Supply of energy is, therefore, far less than the actual demand. Since the potential energy crises are imminent, it is necessary that the world act immediately and vigorously to promote energy efficiency.

In factories, energy conservation is actually aimed at realizing both improvement in productivity and saving energy. Various types of actions are adopted for saving energy.

For air-conditioning facilities, the use of natural ventilation is desirable, as well as solar or terrestrial heat, in addition to better heat insulation technologies in factories and buildings.

For lighting, increased use of natural lighting and lighting using light-emitting diodes (LEDs) is expected as energy saving actions. Especially, improvements in white LED devices for general lighting purpose are under way. White-light LED lamps have longer life expectancy and higher efficiency (the same light for less electricity) than most other lighting. Because of the small size of LEDs, controlling of the spatial distribution of illumination is extremely flexible, and the light output and spatial distribution of a LED array can be controlled with no efficiency loss.

Moreover, lower loss and higher efficiency of power transformers and electric motors are increasedly emphasized in industrial field. The losses of a power transformer are classified in Figure 23.1. Transformers in factories work throughout the day, and it may be quite rare for them to reach the full rated load. Therefore, the most important point is to reduce the no-load losses caused by hysteresis and eddy current losses in the iron core. It is reported that the losses can be reduced down to 50% and 33% under a load factor of 28% and 50%,

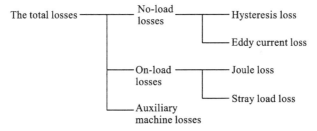

Figure 23.1 Classification of losses in power transformers

respectively, with super amorphous and oil immersion transformers.

One of the quickest ways for industries and utilities to lower energy consumption and therefore reduce their bills is to employ high efficiency motors. Hundreds of millions of electric motors driving machines, compressors, fans, pumps or conveyors in virtually every sector account for about 67 percent of all the electricity that industry uses. Suitable inverters are combined with the electric motors to control their speed. Replacement of ordinary electric motors with smaller variable speed motors, matching output to actual load, can save electricity, avoid pollution and offer economic benefits. In case of a servomotor system, the position of the rotor is monitored and controlled continuously. The rotation speed and direction are able to change quickly and smoothly. In a servomotor system, power is consumed only by the rotating motion. In addition, no power is required while the system is in the standby mode or idling, unless it is required for the holding action. Therefore, a servomotor system will reduce the power consumption substantially, thereby saving energy.

On the other hand, the following technology has been applied for energy saving in various industrial machines and in some types of facilities. About half a century ago, a dynamic braking method was developed for train braking, wherein the kinetic energy of a moving train is used to generate electric current by the locomotive traction motors. Now, the regenerating brake system is used to store electric energy in storage batteries or capacitors, which is generated from the kinetic energy of the machine under stopping or slowing down operation. It is reported that the downward potential energy of a cargo elevator can be converted into regenerated electric energy and stored in Ni-H storage batteries, thereby reducing the consumption of energy by the elevator by over 20%.

The greatest industrial energy savings, though, frequently occur in improving the efficiency of industrial processes themselves, e. g. , using continuous casting of steel and utilizing waste products for electricity and heat generation, as is often done in paper, lumber and plywood manufacturing in the United States.

Utilization of the waste heat from electricity generation for industrial or district heating purposes converts as much as 90% of fuel input into useful energy, compared to 30%~35% for a conventional power plant, thus saving significant amounts of fuel and avoiding pollution. Conversely, some manufacturing facilities that produce substantial amounts of high temperature fluid or steam wastes have used this waste heat for electricity production. Combined heat and power (CHP) is most efficient when heat can be used on-site or very close to it. Overall efficiency is reduced when the heat must be transported over longer distances. This requires heavily insulated pipes, which are expensive and inefficient; whereas electricity can be transmitted along a comparatively simple wire, and over much longer distances for the same energy loss.

Roughly 81 GW of CHP was installed in the US as of 2004, providing about 12% of total electricity production. Europe is far ahead of the US in CHP installation, exceeding 30% in the Scandinavian countries and being widely used in the climate strategies of the UK, Denmark, Sweden, the Netherlands and Germany . There is enormous potential to expand the use of CHP. For example, the US chemicals industry uses only about 30% of its CHP potential. All US conventional power plants together convert only one third of their fuel into electricity, thus wasting two thirds as waste heat, which is equivalent to the total energy use of Japan. Fully adopting this one innovation would profitably reduce total US carbon dioxide emissions by about 23%. Selling waste heat from industrial processes to others within affordable distances could cost-effectively save about 30% of US industrial energy.

New Words and Expressions

1. imminent   即将来临的,逼近的
2. International Energy Agency
   国际能源机构
3. ventilation   通风(装置)
4. terrestrial   陆地的,地球的
5. insulation   绝缘,绝热,隔离
6. light emitting diode(LED)   发光二极管
7. rated   额定的,定价的
8. hysteresis   滞后作用,磁滞现象
9. eddy current   涡(电)流
10. amorphous   非晶体的
11. oil immersion transformer   油浸式变压器
12. inverter   变换器
13. servomotor   伺服电动机
14. locomotive traction motor
    机车牵引电动机
15. brake   刹车
16. capacitor   电容器
17. combined heat and power(CHP)
    热电联产

Notes

(1) One of the quickest ways for industries and utilities to lower energy consumption and therefore reduce their bills is to employ high efficiency motors. Hundreds of millions of electric motors driving machines, compressors, fans, pumps or conveyors in virtually every sector account for about 67 percent of all the electricity that industry uses.

对工业和公共事业部门来说,要降低能量消耗、减少费用支出,最快捷的方法之一是采用高效率的电动机。事实上在每个部门中,采用了成千上万的电动机以驱动机械、压缩机、风扇、泵或者带式运输机等,其用电量约占所有工业用电的67%。

(2) Utilization of the waste heat from electricity generation for industrial or district heating purposes converts as much as 90% of fuel input into useful energy, compared to 30% ~35% for a conventional power plant, thus saving significant amounts of fuel and avoiding pollution.

相对于常规的发电厂30%~50%的燃料利用率,利用发电产生的废热为工业或辖区供热,将90%的输入燃料转换成有用的能量,能节约大量的燃料,避免污染。

❖ Interview Skills

面试技巧(一)

面试=技能+行为。

面试分数一般分为两部分,技能分数与行为分数。技能方面重点考查应聘者专业知识和相关经验。行为方面重点考查一个人是用什么样的行为完成工作的。也许你可以完成这项任务,但是怎样完成的,这个过程也很重要。也就是要考察一个人的综合能力。只有将技能与分数结合,才能真正衡量出面试者是否适合某一职位需求,是否和公司企业文化匹配。下面以应聘机械工程师为例来说明面试过程。

对机械工程师来说,工作经验是非常重要的。如果没有工作经验,也可以将以前设计过的作品告诉招聘者。

(一) 基本句型表达(basic expressions)

(1) Why did you choose to major in mechanical engineering?

你为什么选择机械工程专业呢?

(2) The most important factor is I like tinkering with machines.

最重要的一个因素就是我非常喜欢摆弄机械。

(3) I am very interested in the field your company is in.

我对贵公司所从事的这个领域非常感兴趣。

(4) How do you see your career development?

你如何看待你的事业发展呢?

(5) Why do you think you are qualified for this position?

你为什么认为你适合这份工作呢?

(6) Can you tell me about one of your designs?

你能介绍一项你的设计成果吗?

(二) 会话(conversations)

(A=Applicant I=Interviewer)

Dialogue 1

I:Come in,please.

请进。

A:Good afternoon,Mrs Smith.

下午好,史密斯女士。

I:Good afternoon. Have a seat,please. You are Mr. Sun?

下午好,请坐。你是孙先生吧?

A:Thank you. Yes,I am Sunlin.

谢谢。是的,我是孙林。

I:I have read your resume. I know you have worked for 3 years. Why did you choose to major in mechanical engineering?

我看过你的简历,我知道你已经工作3年了。为什么你选择了机械工程专业呢?

A:Many factors led me to major in mechanical engineering. The most important factor is

I like tinkering with machines.

许多因素导致我选择了机械工程这个专业,但最重要的一个因素就是我非常喜欢摆弄机械。

I: What are you primarily interested in about mechanical engineering?

关于机械工程,你最感兴趣的是什么?

A: I like designing products, and one of my designs received an award. Moreover, I am familiar with CAD. But I can do any mechanic well if I am employed.

我喜欢设计产品,我的一份设计作品还得过奖。而且,我非常熟悉 CAD。但如果我被录用,我什么都能做好。

I: Why did you decide to apply for this position?

你为什么要应聘这份工作呢?

A: Your company has a very good reputation, and I am very interested in the field your company is in.

贵公司的声誉很好,而且我对贵公司所从事的这个领域非常感兴趣。

I: What do you think determines an employee's progress in a company such as ours?

你认为在我们这样的公司里是什么决定着一个员工的发展?

A: Interpersonal and technical skills.

人际关系的技巧和技术技能。

I: We have several applicants for this position. Why do you think you are the person we should choose?

我们有几个应聘者,为什么你认为你就是我们应该选择的那个呢?

A: I have the abilities, qualities and experience that you requested in your job advert, for example I have three years experience in designing products and I got leadership experience while serving the college student union as president.

我具备你们招聘广告上所要求的能力、品质以及工作经验,比如我有三年的产品设计经验,而且我有领导才能,在大学时担任过学生会的主席。

I: That sounds very good. How do you see your career development?

听起来非常不错。你如何看待你的事业发展呢?

A: After a few years of gaining experience in the company and furthering my professional qualifications I'll like to put my experience and skills to use in management. I want to become a supervisor in your R&D department.

在公司具有了几年的经验以及提高专业能力之后,我想把我的经验和技能运用到管理方面,我想成为贵公司研发部的主管。

I: Have you anything to ask about the job?

对这个工作,你有什么问题要问吗?

A: Yes. Do you offer any opportunities for further study?

有。你们提供进修的机会吗?

I: Yes. If you undertake additional courses, provided these are approved, and you complete them successfully, you can claim back part, quite a large part, 75% of the costs you incurred. Not just the fees, traveling and other expenses too.

提供,如果你被批准学习额外的课程,而你又顺利学完的话,你就可以申请补助,多达你所花费的 75%。不仅仅是课程费,还包括差旅费以及其他费用。

A:That's fine.

太好了。

I:Anything else?

还有问题吗?

A:No.

没有了。

I:Well, thank you very much, Mr. Sun. I will let you know the result of the interview as soon as possible. Goodbye.

那好,谢谢你,孙先生。我会尽快告诉你面试结果的。再见。

A:Thank you, Mrs. Smith. I do hope the answer will be favorable. Goodbye.

谢谢你,史密斯女士。我希望是好的。再见。

❖ Reading Material

Energy Saving Factory Automation Equipment and the Environment

The increasing cost of oil and concern for the environment have accelerated the introduction of motor-driven solutions into the industrial field, and therefore the market for motor drive systems is expanding rapidly. Electric motors are the workhorses of industry, driving machines, compressors, fans, pumps and conveyors in virtually all industrial sectors. Every year, several more millions of motors are added. Electric motor driven equipment account for 64 percent of the electricity consumed in the U. S. industrial sector. These systems consume 290 billion kW · h per year. The energy cost needed to operate machinery throughout its useful life can easily exceed the original equipment cost. The application of energy efficient motor systems can greatly reduce the overall environmental impact and cost to operate this equipment.

Reducing the energy consumption of electric motors is broken into two categories. Direct energy consumption, which is the energy consumed while performing work, and indirect consumption, the fixed energy consumed regardless of the operational state. Reducing the direct energy consumption requires enhancing the system efficiency, while reduction in indirect usage requires reductions in the cycle, ancillary and latency times.

1. Reduction in direct energy consumption

Direct energy consumption is the energy consumed during the production cycle. It is the energy needed to do work. Improvements in this area require improvements to the drive systems and machinery selected. In industrial machines the selection of energy saving equipment is typically done during the design phase but in many cases can be retrofitted later onto aging machinery.

One of the most direct ways to reduce energy consumption is to only run motors while the operation is in cycle. Industrial machines using hydraulics or mechanical systems driven by induction motors require stored energy to assist them in performing the work. Since these systems cannot switch on and off quickly and are not easily controlled, they are forced to remain on often at constant speed. The large power output of a synchronous motor allows for the motor to be idle when not in process significantly reducing the power consumption. Replacing inefficient induction motors or hydraulic systems with modern permanent magnet synchronous motors is often the first step.

Implementation of power source regeneration is another excellent means of reducing the power consumption of an electric motor system. During deceleration of an electric motor it will act as generator and energy is put back into the system. In a conventional system that energy is sent to a discharge resistor. A discharge resistor dumps that electrical energy in the form of heat. That waste heat is non-recoverable. By contrast power source regeneration returns electricity to the supply line to be used by other equipment. The electrical energy is recovered and waste heat is greatly reduced.

Power source regeneration requires a drive system with an intelligent power module that can sense the flow of current and switch accordingly. When used the effects are significant. Implementing a servo driven system utilizing drive amplifiers with power source regeneration can see savings of 30% to 40% in electrical power consumption.

Pulse width modulated (PWM) amplifiers provide an excellent control method for electric motors in automation equipment. Rapid acceleration with precise speed and current control is possible but there are always switching losses when using a PWM drive system. These losses translate into waste heat generation. Using the latest generation of power devices will increase controllability and reduce the heat loss. Using fast switching transistors and an increased PWM rate will lower iron losses in both permanent magnet and induction motors. Reduction in heat and other losses within the motors by using the latest power devices and control systems positively contribute to increased performance and reduced energy consumption.

2. Reduction in indirect energy consumption

Indirect consumption is the often overlooked energy costs associated with the lost of production. Optimizing cycle times by improving the acceleration and increasing the movement rates will result in increased productivity but will also reduce the energy consumed. Reducing the time required for a motor to accelerate can provide a vast improvement in the process cycle time. In order to reduce the cycle time the motor core shape must be optimized, inertia reduced and motor control matched closely to the needs of the system. Using permanent magnet synchronous motors provide high power density and excellent acceleration. They are best implemented in high torque, high acceleration but lower ratational speed applications.

Fixed energy consumption includes all the fan motors, pumps, lights, etc. involved in the process. Reduction of the fixed value requires reduction in the time or amount of machines needed for production. Reducing the acceleration time and increased per machine speeds fills these requirements.

The cycle time is reduced by rapid acceleration and often times a smaller motor size is required due to its wide output range enabling further environmental improvements in reduced energy and material usage.

Production and the environment do not need to be at odds. Advancements in electric motor design and the associated drive system in modern automation equipment can be extremely energy efficient. Much of the time the most energy efficient machine will also have the highest performance but there is normally an upfront cost associated with the performance and efficiency. That added cost related with selecting energy efficient equipment may be easily returned as lowered energy costs, high performance and reliability. Too often inefficient machines are selected based on purchase price without concern for the lifecycle cost. The lifecycle cost is the total cost for purchasing, installing, operating, maintaining and disposing of an item of machinery. The purchase price of an electric motor, for instance, is just 1~2 percent of what the owner will spend on energy to run the equipment over its lifetime(see Figure 23.2). As energy prices continue to increase the total lifecycle cost of inefficient machinery will come to the surface. Environmental laws continue to strengthen; disposal costs will increase and recycling of the finite resources gain importance. It is time to look closely at equipment efficiency and environmental impact as part of the decision making process.

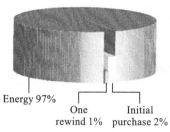

Figure 23.2 Lifecycle cost analysis of motor operations over the motor lifetime

New Words and Expressions

1. retrofit 改型,改进,式样翻新
2. inertia 惯性
3. hydraulics 液压的,水压的
4. torque 扭矩
5. synchronous motor 同步电动机
6. pulse width modulated (PWM)amplifier 脉冲宽度调制放大器
7. permanent magnet motor 永磁电动机
8. induction motor 感应电动机
9. RPM 每分钟转数,转/分
10. implementation 执行,履行,落实
11. acceleration time 加速时间
12. deceleration 减速
13. reliability 可靠性
14. discharge resistor 卸载电阻器,放电电阻器
15. motor core 电动机磁芯
16. dump 倾倒,卸料
17. iron loss 铁损
18. transistor 晶体管
19. generator 发电机,发生器
20. rewind 重绕,重卷
21. at odds 争执,不一致

Lesson 24 Product Control and Quality Assurance

Text

Product control for short is a process for maintaining proper standards in manufacturing. It may include whatever actions a business deems necessary to provide for the control and verification of certain characteristics of a product. The basic goal of product control is to ensure that the products or processes provided meet specific requirements and are dependable and satisfactory. This approach places an emphasis on three aspects:

• Elements such as controls, job management, defined and well managed processes, performance and integrity criteria, and identification of records;

• Competence, such as knowledge, skills, experience, and qualifications;

• Soft elements, such as personnel integrity, confidence, organizational culture, motivation, team spirit, and quality relationships.

Controls include product inspection, where every product is examined visually, using a stereo microscope for fine detail before the product is sold into the external market. Inspectors will provide the lists and descriptions of unacceptable product defects such as cracks or surface blemishes for example. The quality of the outputs is at risk if any of these three aspects is deficient in any way.

Product quality control emphasizes testing of products to uncover defects and reporting to management who makes decision to allow or deny product release, whereas quality assurance attempts to improve and stabilize production (and associated processes) to avoid, or at least to minimize issues which led to the defect(s) in the first place. So, what is quality assurance? Quality assurance (QA) refers to the planned and systematic activities implemented in a quality system so that quality requirements for a product or service will be fulfilled. It is the systematic measurement, comparison with a standard, monitoring of processes and an associated feedback loop that confers error prevention. This can be contrasted with quality control(QC), which is focused on process outputs. You can get the difference between QC and QA in Figure 24.1.

Two principles included in QA are: "Fit for purpose", the product should be suitable for the intended purpose; and "Right first time", mistakes should be eliminated. QA includes supervise of the quality of raw materials, assemblies, products and components, services related to production, management, production and inspection processes.

Suitable quality is determined by users, clients or customers, not by society in general. It is not related to cost, so adjectives or descriptors like "high" and "poor" are not applicable. For example, a low priced product may be viewed as highly-qualified because it is disposable

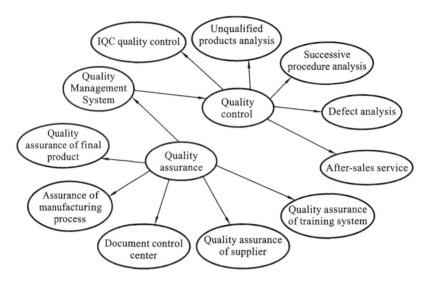

Figure 24.1 The constructs of quality control and quality assurance

where another may be viewed as having poor quality because it is not disposable.

There are many forms of QA processes, of varying scope and depth. The application of a particular process is often customized to the production process.

A typical process may include:
- tests of previous articles;
- plans to improve;
- designs to include improvements and requirements;
- manufacturing with improvements;
- review of new item and improvements;
- tests of the new item.

Quality assurance has gone through three phases.

1) Initial efforts to control the quality of production

During the Middle Ages, guilds adopted responsibility for quality control of their members, setting and maintaining certain standards for guild membership.

Royal governments were interested in quality control as customers. For this reason, King John of England appointed William Wrotham to report about the construction and repair of ships. Centuries later, Samuel Pepys, Secretary to the British Admiralty, appointed multiple such overseers.

Prior to the extensive division of labor and mechanization resulting from the Industrial Revolution, it was possible for workers to control the quality of their own products. The Industrial Revolution led to a system in which large groups of people performing a specialized type of work were grouped together under the supervision of a foreman who was appointed to control the quality of work manufactured.

2) Wartime production

At the time of the First World War, manufacturing processes typically became more complex with larger numbers of workers being supervised. This period saw the widespread introduction of mass production and piece work, which created problems as workmen could now earn more money by the production of extra products, which in turn occasionally led to poor quality workmanship being passed on to the assembly lines. To counter bad workmanship, full time inspectors were introduced to identify, quarantine and ideally correct product quality failures. Quality control by inspection in the 1920s and 1930s led to the growth of quality inspection functions, separately organised from production and large enough to be headed by superintendents.

The systematic approach to quality started in industrial manufacturing during the 1930s, mostly in the USA, when some attention was given to the cost of scrap and rework. The impact of mass production required during the Second World War made it necessary to introduce an improved form of quality control known as statistical quality control, or SQC. Some of the initial work for SQC is credited to Walter A. Shewhart of Bell Labs, starting with his famous one-page memorandum of 1924.

SQC includes the concept that every production piece cannot be fully inspected into acceptable and non acceptable batches. By extending the inspection phase and making inspection organizations more efficient, it provides inspectors with control tools such as sampling and control charts, even where 100 per cent inspection is not practicable. Standard statistical techniques allow the producer to sample and test a certain proportion of the products for quality to achieve the desired level of confidence in the quality of the entire batch or production run.

3) Postwar

In the period following World War II, many countries' manufacturing capabilities that had been destroyed during the war were rebuilt. General Douglas MacArthur oversaw the rebuilding of Japan. During this time, General MacArthur involved two key individuals in the development of modern quality concepts: W. Edwards Deming and Joseph Juran. Both individuals promoted the collaborative concepts of quality to Japanese business and technical groups, and these groups utilized these concepts in the redevelopment of the Japanese economy.

Although there were many individuals trying to lead United States industries towards a more comprehensive approach to quality, the U. S. continued to apply the Quality Control (QC) concepts of inspection and sampling to remove defective product from production lines, essentially ignoring advances in QA for decades.

New Words and Expressions

1. criteria 标准,准则
2. qualification 资格,授权;条件,限制;合格证书
3. stereo microscope 体视显微镜
4. surface blemishes 表面瑕疵
5. guild membership 同业公会,行业联盟
6. statistical quality control 统计质量控制

Notes

(1) Controls include product inspection, where every product is examined visually, using a stereo microscope for fine detail before the product is sold into the external market.

生产控制包括成品检验,在成品检验中要检验每个产品的外观,并且在产品在市场上出售之前通常要使用立体显微镜做细致入微的检查。

(2) Product quality control emphasizes testing of products to uncover defects and reporting to management who makes decision to allow or deny product release, whereas quality assurance attempts to improve and stabilize production (and associated processes) to avoid, or at least to minimize issues which led to the defect(s) in the first place.

产品质量控制强调产品测试以发现缺陷并向决定产品发售与否的管理者报告,而质量保证的目的是提高和稳定质量(结合工艺),从而在一开始就避免,至少最小化导致缺陷的问题。

(3) By extending the inspection phase and making inspection organizations more efficient, it provides inspectors with control tools such as sampling and control charts, even where 100 per cent inspection is not practicable.

它通过延长检测阶段和提高检测机构的效率,提供给检测者诸如采样和控制图表等控制工具,但也不可能实现百分之百的检测。

❖ Interview Skills

面试技巧(二)

Dialogue 2

I: Why do you think you are qualified for this position?

你为什么认为你适合这份工作?

A: I have four years study in the Department of Mechanical Engineering and it has given me a solid theory foundation. Moreover, I have worked at CBA Company for 4 years and got a lot of practical experiences.

我在机械工程系学了4年,这为我的理论水平打下了坚实的基础。而且我在CBA公司工作了4年,获得了丰富的实际经验。

I: Great. Then what is your technical post title now?

太好了。那你现在的技术职称是什么?

A: I am a senior mechanical design engineer.

我现在是高级机械设计工程师。

I: Do you take the original certificate with you?

你把证书原件带来了吗?

A: Yes. Here it is.

带了,给你。

I: Can you briefly tell me about one of your designs?

你能简要介绍一项你的设计成果吗?

A: Of course, I designed a more powerful gasoline engine, which greatly increased the speed of limousines.

当然可以了。我曾经设计了一个功能非常强大的汽油发动机,它显著提高了轿车的速度。

应聘者也应该积极提出问题,比如"Do you offer any opportunities to further study?"这就让招聘者觉得你有积极进取的精神,给他们留下好的印象。

(三) 词汇和短语

tinker with 摆弄
reputation 声誉,名声
supervisor 监督者,主管人
R&D (Research and Development) 研发
leadership 领导才能
approved 经认准的,被认可的
expense 费用,开支
favorable 有利的,顺利的
foundation 基础
technical post title 技术职称
engine 引擎,发动机
limousine 大轿车

(四) 注解

(1) I am familiar with CAD. 我熟悉CAD。
I am accomplished in CAD. 我能熟练使用CAD。
I am well up in CAD. 我精通CAD。
I have mastered CAD. 我熟练掌握了CAD。

(2) What do you think determines an employee's progress(development, advance) in a company such as ours?
在像我们这样的公司,你认为决定员工进步的是什么?

(3) I am a senior mechanical design engineer. 我是高级机械设计工程师。
I am an advertisement designer. 我是广告设计师。
I am an accountant. 我是一名会计。

(4) Can you tell me briefly (concisely, in short) about one of your designs?
你能简单介绍一件你的设计作品吗?

❖ Reading Material

Six Sigma

Six sigma is a business management strategy, originally developed by Motorola in 1986. Six sigma became well known after Jack Welch made it a central focus of his business strategy at General Electric in 1995, and today it is widely used in many sectors of industry.

Six sigma seeks to improve the quality of process outputs by identifying and removing the causes of defects (errors) and minimizing variability in manufacturing and business processes. It uses a set of quality management methods, including statistical methods, and creates a special infrastructure of people within the organization ("Black Belts", "Green Belts", etc.) who are experts in these methods. Each six sigma project carried out within an organization follows a defined sequence of steps and has quantified financial targets (cost reduction and/or profit increase).

The term six sigma originated from terminology associated with manufacturing, specifically terms associated with statistical modeling of manufacturing processes. The maturity of a manufacturing process can be described by a sigma rating indicating its yield, or the percentage of defect-free products it creates. A six sigma process is one in which 99.99966% of the products manufactured are statistically expected to be free of defects (3.4 defects per million). Motorola set a goal of "six sigma" for all of its manufacturing operations, and this goal became a byword for the management and engineering practices used to achieve it.

New Words and Expressions

1. strategy 策略,战略
2. variability 变化性;易变;变化的倾向
3. sequence 数列,序列;顺序;连续;片断插曲

Lesson 25 Agricultural Equipment

Text

Agricultural equipment is equipment used in the operation of an agricultural area or farm. It has a long history and a wide scope.

The first person to turn farming from the hunting and gathering lifestyle probably did so by using his bare hands, and perhaps some sticks or stones. Tools such as knives, scythes, and wooden plows were eventually developed, and had dominated agriculture for thousands of years.

With the coming of the Industrial Revolution and the development of more complicated machines, farming methods took a great leap forward. Instead of harvesting grain by hand with a sharp blade, wheeled machines cut a continuous swathe. Instead of threshing the grain by beating it with sticks, threshing machines separated the seeds from the heads and stalks.

Power for agricultural machinery was originally supplied by horses or other domesticated animals. With the invention of steam power came the portable engine, and later the traction engine, a multipurpose mobile energy source that was the ground-crawling cousin to the steam locomotive. The early traction engine styles look like in Figure 25.1. Agricultural steam engines took over the heavy pulling work of horses, and were also equipped with a pulley that could power stationary machines via the use of a long belt. The steam-powered machines were low-powered by today's standards but, because of their size and their low gear ratios, they could provide a large drawbar pull. Their slow speed led farmers to comment that tractors had two speeds: "slow, and darn slow."

(a) 1882 Harrison Machine Works steam-powered tractor   (b) A very early, hand-buit gasoline powerd Tractor

Figure 25.1 Some style of early tractor

The internal combustion engine, first the petrol engine, and later diesel engines, became the main source of power for the next generation of tractors. These engines also contributed to the development of the self-propelled, combined harvester and thresher, or combine harvester (also shortened to "combine"). Instead of cutting the grain stalks and transporting them to a stationary threshing machine, these combines cut, threshed, and separated the grain while moving continuously through the field.

There're great varieties of AE. In this text, we will introduce some to you.

The farm tractor is used for pulling or pushing agricultural machinery or trailers, for plowing, tilling, disking, harrowing, planting, and similar tasks. A variety of specialty farm tractors have been developed for particular uses. These include "row crop" tractors with adjustable tread width to allow the tractor to pass down rows of corn, tomatoes or other crops without crushing the plants, "wheatland" or "standard" tractors with non-adjustable fixed wheels and a lower center of gravity for plowing and other heavy field work for broadcast crops, and "high crop" tractors with adjustable tread and increased ground clearance, often used in the cultivation of cotton and other high-growing row crop plant operations, and "utility tractors", typically smaller tractors with a low center of gravity and short turning radius, used for general purposes around the farmstead. We can see some modern tractors in Figure 25.2.

(a) A large, modern John Deerd model 9400 four-wheel drive tractor

(b) A modern steerable all-tracked power unit planting wheat in North Dakota

Figure 25.2 Modern tractor

A cultivator is any of several types of farm implement used for secondary tillage. One sense of the name refers to frames with teeth (also called shanks) that pierce the soil as they are dragged through it linearly. Another sense refers to machines that use rotary motion of disks or teeth to accomplish a similar result. The rotary tiller is a principal example. Cultivators stir and pulverize the soil, either before planting (to aerate the soil and prepare a smooth, loose seedbed) or after the crop has begun growing (to kill weeds — controlled disturbance of the topsoil close to the crop plants kills the surrounding weeds by uprooting them, burying their leaves to disrupt their photosynthesis, or a combination of both). Unlike a harrow, which disturbs the entire surface of the soil, cultivators are designed to disturb the soil in careful patterns, sparing the crop plants but disrupting the weeds. Cultivators are usually either self-propelled or drawn as an attachment behind either a two-wheel tractor or four-wheel tractor. For two-wheel tractors they are usually rigidly fixed and powered via couplings to the tractors' transmission. For four-wheel tractors they are usually attached by means of a three-point hitch and driven by a power take-off (PTO). Drawbar hookup is also still commonly used worldwide. Draft-animal power is sometimes still used today, being somewhat common in developing nations although rare in more industrialized economies.

The plough or plow is a tool (or machine) used in farming for initial cultivation of soil in preparation for sowing seed or planting. It has been a basic instrument for most of recorded history, and represents one of the major advances in agriculture. By the pictures of earlier and modern plough in Figure 25.3, we can know the development of plough.

(a) Chinese iron plow   (b) A four-furrow reversible plough

Figure 25.3 Earlier and modern plough

The primary purpose of ploughing is to turn over the upper layer of the soil, bringing fresh nutrients to the surface, while burying weeds, the remains of previous crops, and both crop and weed seeds, allowing them to break down. It also aerates the soil, allows it to hold moisture better and provides a seed-free medium for planting an alternate crop. In modern use, a ploughed field is typically left to dry out, and is then harrowed before planting. Modern competitions take place for ploughing enthusiasts like the National Ploughing Championships Ploughs were initially pulled by oxen, and later in many areas by horses (generally draught horses) and mules. In industrialized countries, the first mechanical means of pulling a plough used steam-powered (ploughing engines or steam tractors), but these were gradually superseded by internal-combustion-powered tractors. In the past two decades plough use has reduced in some areas (where soil damage and erosion are problems), in favour of shallower ploughing and other less invasive tillage techniques.

A broadcast seeder, alternately called a broadcast spreader, is a farm implement commonly used for spreading seed, lime, fertilizer, sand, ice melt, etc., and is an alternative to drop spreaders/seeders.

A manure spreader or muck spreader is an agricultural machine used to distribute manure over a field as a fertilizer. A typical (modern) manure spreader consists of a trailer towed behind a tractor with a rotating mechanism driven by the tractor's power take off (PTO). Truck mounted manure spreaders are also common in North America.

Drip irrigation, also known as trickle irrigation or microirrigation or localized irrigation, is an irrigation method which saves water and fertilizer by allowing water to drip slowly to the roots of plants, either onto the soil surface or directly onto the root zone, through a network of valves, pipes, tubing, and emitters. It is done with the help of narrow tubes which deliver water directly to the base of the plant.

The combine harvester, or simply combine, is a machine that harvests grain crops. The typical style of combine harvester look like in Figure 25.4. The name derives from the fact that it combines three separate operations, reaping, threshing, and winnowing, into one single process. Among the crops harvested with a combine are wheat, oats, rye, barley, corn (maize), soybeans and flax (linseed). The waste straw left behind on the field is the remaining dried stems and leaves of the crop with limited nutrients which is either chopped and spread on the field or baled for feed and bedding for livestock.

(a) One style of combine harvester

(b) A John Deere Titan series combine unloading corn

Figure 25.4 Combine harvest

Combine harvesters are one of the most economically important labor saving inventions, enabling a small fraction of the population to be engaged in agriculture.

The basic technology of agricultural machines has changed little in the last century. Though modern harvesters and planters may do a better job or be slightly tweaked from their predecessors, the US $ 250,000 combine of today still cuts, threshes, and separates grain in essentially the same way it has always been done. However, technology is changing the way that humans operate the machines, as computer monitoring systems, GPS locators, and self-steer programs allow the most advanced tractors and implements to be more precise and less wasteful in the use of fuel, seed, or fertilizer. In the foreseeable future, some agricultural machines will be capable of driving themselves, using GPS maps and electronic sensors. Even more esoteric are the new areas of nanotechnology and genetic engineering, where submicroscopic devices and biological processes, respectively, are being used as machines to perform agricultural tasks in unusual new ways.

Agriculture may be one of the oldest professions, but the development and use of machinery has made the job title of farmer a rarity. Instead of every person having to work to provide food for themselves, less than 2% of the U.S. population today works in agriculture, yet that 2% provides considerably more food than the other 98% can eat. It is estimated that at the turn of the 20th century, one farmer in the U.S. could feed 25 people, where today, that ratio is 1∶130 (in a modern grain farm, a single farmer can produce cereal to feed over a thousand people). With continuing advances in agricultural machinery, the role of the farmer will become increasingly specialized and rare.

New Words and Expressions

1. plow　耕，犁耕；犁
2. manure spreader　撒肥机
3. steam power　蒸汽动力
4. drip irrigation　滴管
5. traction engine　牵引机车
6. combine harvester　联合收割机
7. tractor　拖拉机，牵引机
8. cultivator　中耕机；耕耘机
9. broadcast seeder　撒种机

❖ Writing Training

书信的表达方法与写作——书信的格式

书信是一种应用非常广泛的应用文形式之一，是人们使用英语进行交流的一个重要手段。英文书信的种类很多，但可以把它们归为两大类，一类是私人书信（personal letters），另一类是商务书信（business letters），例如申请信、求职信、推荐信、证明信、感谢信等。英文书信的书写有一定的格式，一封完整的书信应包括寄信人的地址和日期、收信人的姓名和地址、称呼、正文、结尾和署名。

一、写信人的地址与日期

英文信函将写信日期写在信的右上角，即写在寄信人的地址下面，这部分为信头。（heading）若是非正式信件（informal letters），只要写信的日期即可。

二、收信人的姓名、地址和称呼

称呼通常以 Dear... 开头，前面不要留空格，称呼后面通常要用逗号，一般公函中可以用 Dear Sir 或 Dear Madam 等。

三、信的正文

信的正文是信的主体部分，可以根据需要适当分段，使文章层次分明、语言通顺、思路清晰。

（一）开头段

1）告知对方你的身份（假如对方不认识你）

例如：

Dear Sir/ Mr. Smith：

　　I was a student at your college, enrolled in Philosophy Department.

　　I am a... at your...

　　I am a... at your college, enrolled in the... course.

　　My name is... , I am...

2）问候收信人（假如他/她是你的朋友）

例如：

Dear Tom：

　　Hello

Hi. How are you?

I hope everything is fine.

How are things going with you?

How are you getting on in / getting along with... ?

3) 解释写信的原因。

(1) 致谢,例如:

Thank you for your letter about studying in Canada.

I am writing to tell you how grateful I am for...

I would like to thank you most sincerely for...

(2) 抱怨,例如:

I am writing to complain about the poor service at your dining-room.

I wish to make a complaint about...

I am writing to draw your attention to...

(3) 致歉,例如:

I am writing to you because I am unable to...

I am terribly sorry that...

I would like to express my apologies for not being able to...

(4) 询问,例如:

I would like to obtain /request/seek/inquire some information about...

I am writing to ask if you can do me a favor.

I would like some detailed information on/about...

四、结束语

结束语的第一个词的首字母须大写,结尾用逗号。朋友之间的信函常用的结束语有 Yours,Your sever;正规信件(formal letters)常用的结束语有 Yours sincerely 等等。

(一) 发出请求

例如:

Please give this matter your immediate attention.

I would very much appreciate it if... as soon as possible.

(二) 提供帮助

例如:

I hope these... will be helpful, and please feel free to contact me for more information.

... will be taking responsibility for you and if you should need any assistance, she/he will be pleased to help you.

(三) 再次表示歉意或感激

例如:

Thank you for your kind assistance.

Please accept my heartfelt thanks and deepest gratitude, now and always.

I am sorry that I cannot... , and trust that you will understand.

In addition, let me apologize for any inconvenience I may have caused.

Once again, I am sorry for any inconvenience cause.

（四）期盼回信

例如：

I look forward to your prompt response.

Looking forward to a prompt reply.

I expect to hear from you very soon.

I hope to receive your reply shortly.

（五）署名

署名写在结束语下面。

❖ Reading Material

Mechanized Agriculture

Mechanized agriculture is the process of using agricultural machinery to mechanize the work of agriculture, greatly increasing farm worker productivity. In modern times, powered machinery has replaced many jobs formerly carried out by men or animals such as oxen, horses and mules.

The history of agriculture contains many examples of tool use, but only in recent time has the high rate of machine use been at such a level.

The first pervasive mechanization of agriculture came with the introduction of the plough, usually powered by animals. It was invented in ancient Mesopotamia.

Current mechanized agriculture includes the use of tractors, trucks, combine harvesters, airplanes (crop dusters), helicopters, and other vehicles. Modern farms even sometimes use computers in conjunction with satellite imagery and GPS guidance to increase yields.

Since the beginning of agriculture threshing was done by hand with a flail, requiring a great deal of labor. The threshing machine, which was invented in 1794 but not widely used for several more decades, simplified the operation and allowed the use of animal power.

Before the invention of the grain cradle (ca. 1790) an able bodied laborer could reap about one quarter acre of wheat in a day using a sickle. It was estimated that for each of Cyrus McCormick's horse pulled reapers (ca. 1830s) freed up five men for military service in the U.S. Civil War. Later innovations included raking and binding machines. By 1890 two men and two horses could cut, rake and bind 20 acres of wheat per day.

In the 1880s the reaper and threshing machine were combined into the combine harvester. These machines required large teams of horses or mules to pull.

Steam power was applied to threshing machines in the late 19th century. There were steam engines that moved around on wheels under their own power for supplying temporary power to stationary threshing machines. These were called road engines, and Henry Ford seeing one as a boy was inspired to build an automobile.

With internal combustion came the first modern tractors in the early 1900s, becoming more popular after the Fordson tractor (ca. 1917). At first reapers and combine harvesters were pulled by tractors, but in the 1930s self powered combines were developed. (Link to a chapter on agricultural mechanization in the 20th Century at reference.)

The horse population in the U. S. began to decline in the 1920s after the conversion of agriculture and transportation to internal combustion. Peak tractor sales in the U. S. were around 1950. In addition to saving labor, this freed up much land previously used for supporting draft animals.

The greatest period of growth in agricultural productivity in the U. S. was from the 1940s to the 1970s, during which time agriculture was benefiting from internal combustion powered tractors and combine harvesters, chemical fertilizers and the green revolution.

New Words and Expressions

1. mechanized   使(过程、工厂等)机械化
2. productivity   生产力,生产率;丰饶,多产
3. pervasive   普遍的,扩大的,渗透的,弥漫的;无处不在地,遍布地
4. combustion   燃烧,氧化;极度的激动;骚动,混乱

Lesson 26 Quality and Environmental Management Systems

Text

Consumers are increasingly demanding high-quality products and services at low prices, and are looking for suppliers that can respond to this demand consistently and reliably. This trend has, in turn, created the need for international conformity and consensus regarding the establishment of methods for quality control, reliability, and safety of a product. In addition to these considerations, equally important concerns regarding the environment and quality of life are now being addressed with new international standards.

1. The ISO 9000 Quality Management Standard

A quality management system (QMS) can be expressed as the organizational structure, procedures, processes and resources needed to implement quality management. Early quality management systems emphasized predictable outcomes of an industrial production line, using simple statistics and random sampling. By the 20th century, labor inputs were typically the most costly inputs in most industrialized societies, so focus shifted to team cooperation and dynamics, especially the early signaling of problems via a continuous improvement cycle. In 1987 the International Organization for Standardization (ISO) published its first set of standards for quality management called ISO 9000. The ISO 9000 series consists of a set of standards and a certification process for companies. It is applicable to all types of companies and have gained global acceptance and has permanently influenced the way manufacturing company conduct business in world trade. By receiving ISO 9000 certification, companies demonstrate that they have met the standards specified by the ISO. In many industries ISO certification has become a requirement for doing business. In December 2000 the first major changes to ISO 9000 were made, introducing the following three new standards:

• ISO 9000: 2000—*Quality Management Systems-Fundamentals and Standards*: Provides the terminology and definitions used in the standards. It is the starting point for understanding the system of standards.

• ISO 9001: 2000—*Quality Management Systems-Requirements*: This is the standard used for the certification of a firm's quality management system. It is used to demonstrate the conformity of quality management systems to meet customer requirements.

• ISO 9004: 2000—*Quality Management Systems-Guidelines for Performance*: Provides guidelines for establishing a quality management system. It focuses not only on meeting customer requirements but also on improving performance.

The ISO 9000 standard is based on eight principles that emphasize the importance of: customer focus; leadership; involvement of people at all levels; process approach; systems and objectives; continual improvement; accurate data and analysis; and supplier relationships.

Principle 1—Customer-focused organization: Organizations depend on their customers and, therefore, should understand current and future customer needs, meet customer requirements and strive to exceed customer expectations.

Principle 2—Leadership: Leaders establish unity of purpose and direction of the organization. They should create and maintain the internal environment in which people can become fully involved in achieving the organization's objectives.

Principle 3—Involvement of people: People at all levels are the essence of an organization, and their full involvement enables their abilities to be used for the organization's benefit.

Principle 4—Process approach: A desired result is achieved more efficiently when related resources and activities are managed as a process.

Principle 5—System approach to management: Identifying, understanding and managing a system of interrelated processes for a given objective improve the organization's effectiveness and efficiency.

Principle 6—Continual improvement: Continual improvement should be a permanent objective of the organization.

Principle 7—Factual approach to decision making: Effective decisions are based on the analysis of data and information.

Principle 8—Mutually beneficial supplier relationships: An organization and its suppliers are interdependent, and a mutually beneficial relationship enhances the ability of both to create value.

To receive the ISO 9000 certification, a company must provide extensive documentation of its quality processes. This includes methods used to monitor quality, methods and frequency of worker training, job descriptions, inspection programs, and statistical process-control tools used. High-quality documentation of all processes is critical. The company is then audited by an ISO 9000 registrar who visits the facility to make sure the company has a well-documented quality management system and the process meets the standards. If the registrar finds that all is in order, certification is received. Once a company is certified, it is registered in an ISO 9000 directory that lists certified companies. The entire process can take 18 to 24 months and can cost anywhere from $10,000 to $30,000. Companies have to be recertified by ISO every three years. One of the shortcomings of ISO 9000 certification is that it focuses only on the process used and conformance to specifications. ISO 9000 certification does not address questions about the product itself and whether it meets customer and market requirements.

2. ISO 14000 Environmental Management Standard

What is an environmental management system(EMS)? How does it work? It goes like this: an environmental management system is "a framework for implementing your environmental policy and objectives and, through recorded evidence, achieving conformance with the standard that can be demonstrated to others." Three formal types of environmental management systems have been developed to date. Perhaps the well known of these standards is ISO 14000.

ISO 14000 is a series of standards developed by the International Organization in September 1996. It concerns the way an organization's activities affect the environment throughout the life of its products. Theses activities may be internal or external to the organization, ranging from production to ultimate disposal of the product after its useful life, and include effects on the environment such as pollution, waste generation and disposal, noise, depletion of natural resources, and energy use. Briefly stated, the ISO 14000 series covers the following standards:

- ISO 14001—*Environmental Management Systems (EMS)*. The formal elements of an environmental management system include environmental policy, planning, implementation, verification, and management review.
- ISO 14004—*General Guidance for Developing and Implementing an EMS.*
- ISO 14010-12—*Environmental Auditing Principles and Guidance.*
- ISO 14031—*Environmental Performance Evaluation Guidance.*
- ISO 14020-24—*Environmental Labeling Guidance* (products).
- ISO 14040-45—*Life-cycle Assessment Principles and Guidance* (mainly products).
- ISO 14050—*Terms and Definitions.*

ISO 14000 standard includes information on all of the elements needed to develop an environmental management system in an organization. It provides the requirements that an organization must conform to in order to obtain third party registration or certification.

To introduce the framework for an EMS, according to ISO 14001 (see Figure 26.1), the first thing is to establish broad environmental goals (environmental policy commitment). Then, the company's activities and processes should be assessed (environmental audit), followed by the environmental implications of its business operations (aspects and impacts

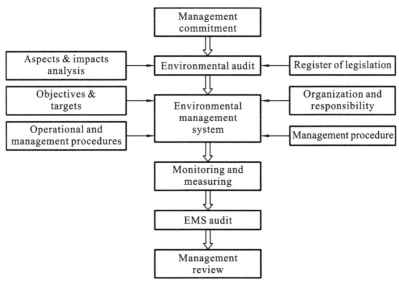

Figure 26.1 **Content of the environmental management plan**

analysis). From this information, it is possible to set environmental improvement standards (objectives and targets) and establish a programmer to achieve the objectives and targets (operational and management procedures). When all of this is in place, the progress made in achieving the objectives and targets (environmental management system audit) may be evaluated and, finally, the system in the light of the findings (management review) can be reviewed. The underlying emphasis of the system is on continuous improvement; this can occur in two ways. First, as objectives and targets are set and achieved, so others take their place and the company continually makes its operations more environmentally friendly. Second, the environmental management system itself improves continuously by becoming more streamlined, e. g. initial procedures may seem cumbersome with unnecessary paperwork for the size of the company. Regular EMS review meetings with employees will assist with the streamlining and make the EMS fit the way a company operates.

With respect to ISO 14000, registration is the formal recognition of an organization's ability to conform to the requirements of an EMS. Organizations may simply declare that their EMS meets the requirements of ISO 14001 ("self-declaration"). However, many organizations choose to have their EMS registered, usually to provide greater assurance to clients and the public, or because regulators and clients require it. A rapidly increasing number of companies in many countries have been obtaining certification for this standard.

3. Relationship between ISO 9000 and ISO 14000 standards

ISO 9000 is primarily concerned with quality management. The definition of "quality" in ISO 9000 refers to all those features of a product or a service which are required by the customer. Quality management means what the organization does to ensure that its products conform to the customer's requirements. The ISO 9000 standard has been developed specifically to address customer requirements and expectations regarding product quality. ISO 14000 is primarily concerned with environmental management. This means what the organization does to eliminate harmful effects on the environment caused by its activities.

Although there may be a significant difference in procedural detail between ISO 9000 and ISO 14000 standards, the ISO 14000 standard uses similar frameworks as ISO 9000 such as document control, management system auditing, operational controls, recordkeeping controls, management policies, audits, training, and corrective and preventive actions. ISO 9000 and ISO 14000 require senior management support and commitment for success, and require organizations to have a system for establishing and reviewing objectives and targets, whether they are quality or environmentally related. Both require organizations to provide on-going management review of the management system and its objectives. If the ISO 9000 is already in place, the introduction of ISO 14000 will be relatively straightforward. Conversely, if there is not any kind of quality management system, the introduction of ISO 14000 first will greatly reduce the time and effort needed to achieve ISO 9000 accreditation in the future.

New Words and Expressions

1. consensus （意见等）一致
2. series 系列
3. implement 贯彻，执行
4. environmental policy commitment 环境政策承诺
5. statistics 统计数据，统计学
6. environmental audit 环境审核
7. random sampling 随机抽样
8. management review 管理调查
9. certification 证明，鉴定，证书
10. registration 登记，注册
11. audit 审计；查账
12. accreditation 委派，鉴定
13. verification 证明，证实，核实

Notes

（1）Consumers are increasingly demanding high-quality products and services at low prices, and are looking for suppliers that can respond to this demand consistently and reliably.

消费者对质量高、价格低的产品和服务的需求在不断增加，而且他们希望供应商能持续、可靠地提供优质产品和服务。

（2）The ISO 9000 standard is based on eight principles that emphasize the importance of: customer focus; leadership; involvement of people at all levels; process approach; systems and objectives; continual improvement; accurate data and analysis; and supplier relationships.

ISO 9000 标准通过八项原则强调以客户为中心、领导作用、全员参与、过程方法、系统管理方法、持续改进、准确记录和分析数据、改善与客户的关系的重要性。

（3）The underlying emphasis of the system is on continuous improvement; this can occur in two ways. First, as objectives and targets are set and achieved, so others take their place and the company continually makes its operations more environmentally friendly. Second, the environmental management system itself improves continuously by becoming more streamlined, e.g. initial procedures may seem cumbersome with unnecessary paperwork for the size of the company.

环境管理标准体系强调的是持续改进，一般通过两种方式实现：首先，当企业预定的环境目标实现以后，要确立新的目标使企业持续以更加环境友好的方式运行；第二，环境管理标准体系也在不断简化和完善，例如，最初制定的标准规程因要求企业准备不必要的文字材料而显得过于烦琐。

（4）Although there may be a significant difference in procedural detail between ISO 9000 and ISO 14000 standards, the ISO 14000 standard uses similar frameworks as ISO 9000 such as document control, management system auditing, operational controls, recordkeeping controls, management policies, audits, training, and corrective and preventive actions.

虽然 ISO 9000 和 ISO 14000 标准的程序内容差别很大，但是两种标准的框架是非常相似的，例如二者都包含文件记录、管理系统审核、运行控制、记录保持控制、管理政策、审核、培训

以及纠正和预防措施。

❖ Writing Training

书信的表达方法与写作——书信中感谢的表达方法

书信中感谢的表达方法有很多,特列举如下。

As soon as I opened your package, I felt that I must sit right down and tell you...
一打开您寄来的包裹,就觉得应该立即写信致谢。

I just wish you could have seen ...
我愿您知道……

You were kind to send a gift ...
承蒙好意,送来礼品……

Your lovely gift came this morning(was waiting for us when)...
厚礼于今早收到(当……时,厚礼正等我们去拿)

Expected gifts are a pleasure to receive, but unexpected remembrances are an even greater joy.
意料中的礼物固然可喜,而意料之外的礼物更使人喜出望外。

You couldn't have given me anything that I wanted (would enjoy) more.
再没有比您的礼物更为我所想要的了(更能使我喜欢了)。

You must be a mindreader.
您一定很会摸透别人的心思。

The wedding gift you sent to us is one of the most beautiful we received. It now occupies the most prominent place on our mantel.
承蒙您所惠赠的结婚礼物是我们收到的最佳的礼物之一,现陈设在壁炉架上最显眼的地方。

I find an ordinary "thank you" entirely inadequate to tell you how much ...
我觉得用"感谢您"这种老生常谈的话,远不能表达我对您是多么……

I don't know when I have had such a delightful(pleasant)(memorable)(enjoyable) weekend as the one...
我不记得曾经像那次一样高兴地(愉快地)(难忘地)(快乐地)度过周末过。

This is to thank you again for your wonderful hospitality and to tell you how much I enjoyed seeing you again.
再次感谢您的盛情款待,并希望再次见到您。

Thank you for one of the most enjoyable visits we have had in many months.
在贵处的参观访问,是我们几个月中最愉快的一次。谨向您表示感谢。

Thank you for one of the most memorable days of my trip.
为了旅途中最值得留念的一天,谨向您表示感谢。

Thank you so much for your generous hospitality.
非常感谢您慷慨的款待。

I hope something will bring you to New York soon so that I can reciprocate your

kindness.
希望不久您能有机会到纽约来,使我能答谢您的盛情。
You must give me the chance to return your kindness when you visit here.
希望您光临我处,使我能答谢您的盛情。
Thank you very much (very, very much) (ever so much) (most sincerely) (indeed) (from the bottom of my heart).
很(非常)(非常非常)(最真诚地)(确实)(衷心)感谢您。
Many thanks for your kind and warm letter.
感谢您友好而热情的来信。
Thanks a million (ever so much).
万分(非常)感谢。
Please accept (I wish to express) my sincere (grateful) (profound) appreciation for …
请接受(致以)诚挚的(衷心的)(深切的)感谢……
I sincerely (deeply) (warmly) appreciate …
我诚挚地(深深地)(热情地)感谢……
I am very sincerely (most) (truly) grateful to you for …
为了……,我非常诚挚地(深深地)(真诚地)感谢您。
There is nothing more important (satisfying) (gratifying) to me than to receive one of your letters.
再也没有比收到您的来信更使我觉得重要(快慰)(感激)了。
Your letters are so much fun (comfort) (entertainment) (company).
您的来信充满了乐趣(给了很大安慰)(带来了欢乐)(使我不感寂寞)。
Your most courteous (considerate) (delightful) letter…
您那彬彬有礼(体贴入微)(令人欣慰)的来信……
I cannot tell you how much your letter delighted (relieved) (amused) (enchanted) me.
我无法告诉您,您的来信使我多么高兴(宽慰)(觉得有趣)(陶醉)。
I love the way you say (put) things in your letters. You make even the smallest incident seem so interesting (important) (charming) (mysterious).
我很欣赏您在信中描述各种事物的手法。您的妙笔使最微小的事情都显得很有趣(重要)(动人)(神秘)。
It was good (fine) (charming) (thoughtful) of you…
承蒙好意(美意)(盛情)(关心)……
It was nice (characteristically thoughtful) (more than kind) of you…
承蒙好意(特别关心)(十二分好意)……
At the outset, I want to thank you for your kindness to me and for your compliments.
首先,我要感谢您对我的友爱和问候。
Believe me, I am truly grateful for…
我确实真诚地感谢……
We were deeply touched by …
……使我们深受感动。

It's generous of you to take so much interest in my work(to give me so much of your time) (to show me so much consideration).

承蒙您对我的工作如此操心(为我花费这么多时间)(对我如此关怀)。

We are indebted to you...

我们感谢你……

❖ Reading Materials

Total Quality Management (TQM)

Total quality management (TQM) is an integrated organizational effort designed to improve quality at every level. TQM includes techniques for achieving efficiency, solving problems, imposing standardization and statistical control, and regulating design, housekeeping, and other aspects of business or production processes.

The concept of quality has existed for many years, its meaning has changed and evolved over time. In the early twentieth century, quality management meant inspecting products to ensure that they met specifications. In the 1940s, during World War II, quality became more statistical in nature. Statistical sampling techniques were used to evaluate quality, and quality control charts were used to monitor the production process. In the 1960s, quality began to be viewed as something that encompassed the entire organization, not only the production process. The meaning of quality changed dramatically in the late 1970s. A new concept of quality was emerging. One result is that quality began to have a strategic meaning. The term used for today's new concept of quality is total quality management or TQM. Table 26.1 presents a timeline of the old and new concepts of quality.

Table 26.1 Timeline showing the differences between old and new concepts of quality

TIME:	Early 1900s 1940s 1960s	1980s and Beyond
FOCUS:	Inspection Statistical Organizational sampling quality focus	Customer driven quality
	Old concept of quality: Inspect for quality after production.	New concept of quality: Build quality into the process. Identify and correct causes of quality problems.

The characteristics of TQM is the focus on identifying root causes of quality problems and correcting them at the source, as opposed to inspecting the product after it has been made. Not only does TQM encompass the entire organization, but it stresses that quality is customer driven. TQM attempts to embed quality in every aspect of the organization. It is concerned with technical aspects of quality as well as the involvement of people in quality, such as customers, company employees, and suppliers. Table 26.2 shows the specific concepts that make up the philosophy of TQM.

Table 26.2　Concepts of the TQM philosophy

Concept	Main Idea
Customer focus	Goal is to identify and meet customer needs
Continuous improvement	A philosophy of never-ending improvement
Employee empowerment	Employees are expected to seek out, identify and correct quality problems
Use of quality tools	Ongoing employee training in the use of quality tools
Product design	Products need to be designed to meet customer expectations
Process management	Quality should be built into the process; sources of quality problems should be identified and corrected
Managing supplier quality	Quality concept must extend to a company's suppliers

Customer Focus　The first, and overriding, feature of TQM is the company's focus on its customers. TQM recognizes that a perfectly produced product has little value if it is not what the customer wants. Therefore, we can say that quality is customer driven. Companies need to continually gather information by means of focus groups, market surveys, and customer interviews in order to stay in tune with what customers want.

Continuous Improvement　Continuous improvement requires that the company continually strive to be better through learning and problem solving. A typical approach that can help companies with continuous improvement is the plan-do-study-act (PDSA) cycle. The PDSA cycle describes the activities a company needs to perform in order to incorporate continuous improvement in its operation. Managers must evaluate the current process and make plans based on any problems they find. The next step in the cycle is implementing the plan (do). During the implementation process managers should document all changes made and collect data for evaluation. The third step is to study the data collected in the previous phase and to see whether the plan is achieving the goals established in the plan phase. The last phase of the cycle is to act on the basis of the results of the first three phases. The best way to accomplish this is to communicate the results to other members in the company and then implement the new procedure if it has been successful. Note that this cycle shows that continuous improvement is a never-ending process.

Employee Empowerment　In TQM, the role of employees is very different from what it was in traditional systems. Workers are empowered to make decisions relative to quality in the production process. They are considered a vital element of the effort to achieve high quality. Their contributions are highly valued, and their suggestions are implemented. In order to perform this function, employees are given continual and extensive training in quality measurement tools. In addition, TQM places great emphasis on teamwork. Using techniques such as brainstorming, discussion, and quality control tools, teams work regularly to correct problems. The contributions of teams are considered vital to the success of the company.

Use of Quality Tools　The TQM provides a wide range of quality control tools such as the following: (1) a cause-and-effect diagram is a chart that identifies potential causes for particular quality problems. They are often called fishbone diagrams because they look like the bones of a fish; (2) a flowchart is a schematic diagram of the sequence of steps involved in an operation or process; (3) a checklist is a list of common defects and the number of observed occurrences of these defects. It is a simple yet effective fact-finding tool that allows the worker to collect specific information regarding the defects observed; (4) a control chart is a very important quality control tool used to evaluate whether a process is operating within expectations relative to some measured; (5) a scatter diagram is a graph that show how two variables are related to one another. They are particularly useful in detecting the amount of correlation, or the degree of linear relationship, between two variables; (6) Pareto analysis is a technique used to identify quality problems based on their degree of importance. In quality management the logic behind Pareto's principle is that most quality problems are a result of only a few causes; and (7) a histogram is a chart that shows the frequency distribution of observed values of a variable. We can see from the plot what type of distribution a particular variable displays, such as whether it has a normal distribution and whether the distribution is symmetrical.

Product Design A critical aspect of building quality into a product is to ensure that the product design meets customer expectations. A useful tool for translating the voice of the customer into specific technical requirements is quality function deployment (QFD). QFD is also useful in enhancing communication between different functions, such as marketing, operations, and engineering. QFD begins by identifying important customer requirements, which typically come from the marketing department. These requirements are numerically scored based on their importance, and scores are translated into specific product characteristics. Evaluations are then made of how the product compares with its main competitors relative to the identified characteristics. Finally, specific goals are set to address the identified problems.

Process Management According to TQM, a quality product comes from a quality process. This means that quality should be built into the process. Quality at the source is the belief that it is far better to uncover the source of quality problems and correct it than to discard defective items after production.

Managing Supplier Quality The philosophy of TQM extends the concept of quality to suppliers and ensures that they engage in the same quality practices. If suppliers meet preset quality standards, materials do not have to be inspected upon arrival. Today, many companies have a representative residing at their supplier's location, thereby involving the supplier in every stage from product design to final production.

As we have seen, total quality management has impacts on every aspect of the organization. Every person and every function is responsible for quality and is affected by poor quality. TQM requires the close cooperation of different functions in order to be successful.

New Words and Expressions

1. embed   把……嵌入，融入
2. symmetrical   对称的，匀称的
3. empower   授权，准许
4. Pareto's principle   帕雷托原理
5. cause-and-effect diagram
   原因与结果分析图表
6. plan-do-study-act (PDSA) cycle
   计划-行动-研究-行动循环
7. quality function deployment (QFD)
   质量功能展开
8. specification   规格，规范，说明书

附录 参考译文

第 1 课 工 程 制 图

从视觉考虑,图 1.1 所示的洗衣机和机械手以照片的形式被清晰地表达出来了。但是仅仅依靠照片不能将这些产品制造出来。制造产品时需要有零件图,并清晰画出每个零件的确切形状和标注出其尺寸、材料。另外还需要产品的装配图,该图注明所有零件的装配关系和组装技术要求等。

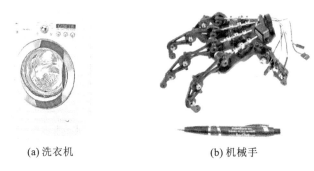

(a) 洗衣机　　　　　　　(b) 机械手

图 1.1　典型机械产品示意图

工程制图是工程师设计产品并给出产品加工说明的一种方法。工程图样可以手绘也可以利用计算机软件绘制,在图样中要给出加工零件的尺寸、装配要求、需要的材料及其他相关信息。工程图样是具有法律效力的文件,其表达必须准确且符合标准。

在现代制造业中,有几种类型的图样可以使用。但是,标准的工程图样是多视图图样。工程图样一般含有 2~3 个视图(主视图、俯视图和侧视图),每个视图均为所绘零件的正投影(见图 1.2①)。在铅直平面(x-z)中的投影是指定的,称为主视图。在水平面(x-y)上的投影沿 x

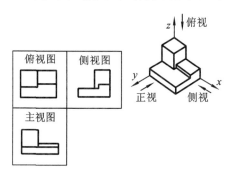

图 1.2　零件多视图表达

---

① 注:该图采用的是第三角投影画法,我国一般采用第一角投影画法。

轴顺时针旋转 90°即为俯视图。在侧立面（y-z）上的投影沿 z 轴顺时针旋转 90°即为侧视图。

线型和规范 图 1.3 给出了图样绘制中使用的各种线形和规范用法。图样绘制的质量取决于线型（包括字体）黑度和均匀度。这里说的线型仅仅是指线的规范形式，在任何情况下，每种线都应该是不透明的，并且粗细均匀。上述线型和规范是指用于工程图样的而不是用于它图图表，如电路图等。

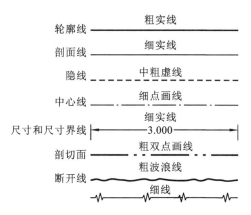

图 1.3 工程制图常用线型

比例尺 工程图样一般采用 1∶1 比例绘制，除非零件尺寸过大画不下，或过小、过于复杂需要放大才能表达清楚。当某个大型零件的主要视图被缩小比例画出时，表达零件中某些细节的局部视图就要尽可能按 1∶1 比例画出。为了清晰表达某个零件，而将零件放大画出时，就不需要画真实尺寸的视图了。工程图样常用的比例为 1∶1，缩小的比例有 1∶2，1∶4，1∶10 和 1∶20，放大的比例有 2∶1，4∶1，10∶1，20∶1。

草图 草图是徒手快速绘制的不作为最终图样使用的图样。通常，草图绘制是将设计思路快速记录下来以备后用的一种方式。徒手绘制草图有利于从设计者到制图员、从车间管理者到操作者的整个工程过程的交流。草图一般要包含绘制正式图样所需的所有设计信息。草图可以附有图解或说明，以便于工程技术人员之间交流。

零件图 零件图是对一个零件完整清晰的表达，要认真选择视图并标注好零件的尺寸。零件图应该包含零件采购、加工所需要的信息，并且确定所用相关的代码和标准。此外，表面粗糙度和加工工艺要求都应该包含在零件图中。零件图中还要包括零件的重量/质量以便查阅。标题栏应该给出制造零件使用的材料以及组装时需要的该零件的数量。有些零件图只注有特殊尺寸和技术要求，这类图样一般是为特殊工种操作人员绘制的，例如模样制作人员、机加工或者焊接工人等。

装配图 装配图中注明了各个零件和各个部件之间的相对位置。典型的装配图包括以下几部分：

- 采用一个或多个视图，包括局部视图和其他辅助视图；
- 用于表达细微结构的放大的局部视图；
- 装配所需的总体尺寸和详细尺寸；
- 装配工艺及技术要求；
- 零件序号；
- 零件列表或物料清单。

装配图按照用途不同分为设计装配图、制造装配图、总装图、安装图和维修图。

剖视图 许多零件的内部结构比较复杂,通过主视图、俯视图、侧视图和示意图不能表达清楚。在很多工程制图样中,采用剖视图表达零件的结构并且标注全部尺寸是必要的。剖视图的特征是带有剖面线(见图 1.4),剖面线符号表示利用想象的剖切面产生剖视图的位置。通常,采用一个剖视图就足以将零件表达清楚,但是对于不规则的零件,就需要有多个剖视图。全剖视图或断面图能展现零件特有的形状,经常取代仅表达外部形状的主视图。为了将零件的内部结构表达得更加清晰,也可将侧视图或俯视图转化成剖视图,或者将主视图表达成剖视图后,根据需要将侧视图或俯视图转化成剖视图。当需要同时表达对称零件的内部和外部结构时,可以采用半剖视图。局部视图主要用来表达零件的内部结构,局部视图的剖面小于半剖视图的。

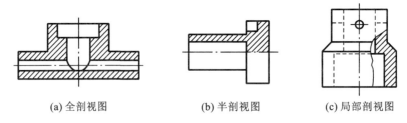

(a) 全剖视图　　　　(b) 半剖视图　　　　(c) 局部剖视图

图 1.4　剖视图的表达

尺寸标注 工程图样表达的内容不仅包括物体的几何学(形状和位置)特性,还包括尺寸和公差特性。尺寸标注体系有多种,最简单的尺寸标注是标注两点间的距离(例如物体的长和宽,或孔的中心位置)。由于互换性制造技术的出现和不断完善,零件图的尺寸都用带有正、负公差或最大、最小极限的尺寸标注。绘制大型零件图时,尺寸单位多用英尺,小型零件的尺寸以英寸为单位。在米制单位中,标注尺寸常用米、厘米或毫米,具体取决于零部件或结构的大小。

分解图 分解图是工程示意图,用来说明各零部件的装配顺序(见图 1.5)。分解图中可以标注零件尺寸和装配尺寸以表明零件之间的关系。尽管分解图不能作为机加工用图,但仍然是机械工艺的重要组成部分。分解图广泛用于产品手册中,方便对机械装置或其他机构进行维修和组装。

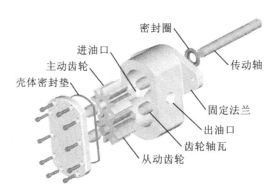

图 1.5　齿轮泵立体分解系统图

原理图 原理图或简图是用符号构成的表示组成一个系统的重要零件、部件、电路(包括它们的相互联系)、设备、流程、工序及方案的图形。例如电路图,它是用来表示电路、设备或系统中各电子元件间的相互关系的功能图。表征电子元件的符号已经被格式化、简化和标准化,能被广泛接受。在机械原理图中,表达机械系统中零件及其相互关系的图形比电路图更复杂,因此机械原理图的特征是不够简洁和标准。工艺流程图常用来说明企业的生产工艺和设备流程。流程图用于表达企业中主要设备的相互关系,图中不包括次要的细节,如详细的管道牌号等。原理图广泛应用于设备维修手册中,帮助使用者了解各部分的相互连接情况,为机械的组装和拆卸提供图解说明。

总体布局图 总体布局图是对零件或装置的总体形式进行说明的图样。通常在产品的研发阶段需要绘制总体布局图。由于总体布局图缺少具体设计细节,只能作为设计者或参与产品初期设计的绘图人员的参考。在大多数情况下,设计人员绘制的总体布局图最终成为其他人员绘制零件图和装配图的第一手资料。

第 2 课 力 学

就最一般的意义而言,力学研究的是物体的受力及力对物体的作用。通常,工程力学适用于根据已知的作用力(也称为载荷)或压力,对物体的加速度和变形(包括弹性变形和塑性变形)进行分析和预测。力学分支学科包括静力学、动力学、材料力学、运动学、连续介质力学。

机械工程师通常会在工程设计或分析阶段使用力学知识。如果工程项目是设计一辆车,为了评估哪里的压力是最大的,可以利用静力学来设计车辆的车架。动力学则可以用在设计汽车的发动机时,以评估活塞和凸轮在发动机运转时内力的变化。材料力学可以用于为车架和发动机选择合适的材料。流体力学可以用于汽车通风系统的设计,或发动机进气系统的设计。

静力学是力学的一个分支,它主要涉及物理系统在受到载荷(力、扭矩/力矩)处于静力平衡时的分析。静力平衡状态即子系统的相对位置不随时间变化,或其中的成分和结构处在一个恒定的速度下。在静力平衡时,系统或者静止,或者其质心以恒定速度移动。

由牛顿第一定律可知,这种情况意味着系统的每个部分的合力和合扭矩(又称力矩)为零。根据这个条件,可以得到内力或压力的数目。合力等于零被称为第一平衡条件,合扭矩等于零被称为第二平衡条件。

在物理学领域,动力学主要研究物体运动及其变化的原因。换句话说,它研究的是作用在物体上的力和力引起物体运动的原因。动力学还研究扭矩对物体运动的影响。这些都与运动学相反。运动学是经典力学的一个分支,主要描述物体的运动,而不考虑导致物体运动的原因。

一般而言,研究人员在动力学方面主要研究一个物理系统是如何发展和随时间变化的,以及导致这些变化的原因。此外,艾萨克·牛顿确立了决定物理动力学的物理规律。动力学可以通过研究系统的力学特性来理解,动力学尤其主要与牛顿第二运动定律相关。然而,对牛顿三定律都需要予以考虑,因为它们在任何给定的观察或实验中都是相互关联的。

在材料力学中,材料的强度是指物体抵抗外力而不破坏的能力。材料强度的研究常常涉及对梁、柱、轴等结构件的应力进行计算的多种方法。这些方法用来预测结构在承受载荷时的反应及其对各种破坏模式的敏感性。运用这些方法时可以考虑材料除屈服强度和极限强度之外的各种性能。例如,研究屈曲破坏要考虑材料的刚度,进而要考虑杨氏模量。

流体力学的研究对象是流体(包括液体、气体和等离子体)和作用在流体上的力。流体力学可以分为:流体静力学,研究流体的静止状态;流体运动学,研究流体的运动状态;流体动力学,研究力对液体运动的影响。流体力学是连续介质力学的一个分支,这个学科并不根据原子提供的信息来考虑物质,也就是说,是从宏观角度而不是从微观角度来研究物质的。流体力学,特别是流体动力学,是一个活跃的研究领域,有许多尚未解决或部分解决的问题。流体力学在数学上是很复杂的。有时最好通过数值方法,通常是使用计算机来解决流体力学问题。一门现代学科,称为计算流体动力学(CFD),专门研究采用这种方法来求解流体力学问题。

运动学的研究对象通常被称为几何运动。为了描述运动,运动学研究点、线和其他几何对象的运动轨迹以及它们的微分性质,如速度与加速度。在天体物理学中,运动学用于描述天体系统的运动,而在机械工程机器人学和生物力学领域,运动学用以描述由连接部分组成的系统(多连杆系统),如发动机、机械臂或人体骨架的运动。

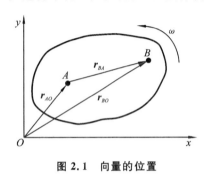

图 2.1 向量的位置

刚体的平面运动如图 2.1 所示。假设刚体中点 A 的角速度为 ω,点 B 为刚体中要研究的任意一点。如图 2.1 所示,向量 r_{AO} 和 r_{BO} 分别表示点 A 和点 B 的相对位置。ω 与点 A 和点 B 的选择无关,因为刚体上的所有直线的转动都是一样的。通过图 2.1 中标出的位置矢量,点 A 和点 B 之间的绝对运动关系可以很容易得到,也就是:

$$r_{BO} = r_{AO} + r_{BA} \tag{1}$$
$$v_B = v_A + v_{BA} = v_A + \omega r_{BA} \tag{2}$$
$$a_B = a_A + a_{BA} = a_A + \alpha r_{BA} - \omega^2 r_{BA} \tag{3}$$

在公式(2)和公式(3)中,v_A、a_A 分别为点 A 的速度和加速度;v_B、a_B 分别为点 B 的速度和加速度;ω 为刚体的角速度,α 为刚体的角加速度。两个向量都垂直于运动平面,所以刚体的转动是平行于坐标轴的。

连续介质力学是力学的一个分支,它涉及模型是连续体而不是离散颗粒的材料的力学性能和运动学的分析。在 19 世纪,法国数学家奥古斯丁·路易斯·柯西第一个用公式表示出了这些模型,但在这一领域的研究一直持续到今天。

第 3 课 工 程 材 料

人类文明的历史发展经历了石器时代、铜器时代、铁器时代、钢铁时代、太空时代(与电子时代是同期的)。每一个时代都以某种材料的出现为标志。当今,用于工业的材料被分为工程材料和非工程材料。工程材料是指用于生产制造并且成为产品的一部分的那些材料;非工程材料是指化学用品、燃料、润滑剂和其他用于生产过程但不会成为产品的组成部分的材料。工程材料构成了技术的基础,不论这项技术属于结构、电子、热能、电化学、环境、生物医学还是其他应用领域。工程材料的主要分类包括金属、聚合物、陶瓷、复合材料等(见图 3.1)。

金属(包括合金)

这类材料由一种或多种金属元素(如 Fe、Al、Cu、Ti、Au 和 Ni)组成,并且通常还含有相对较少的非金属元素(如 C、N 和 O)。金属及其合金中的原子以非常有序的方式进行排列,与陶瓷和聚合物中的原子相比排列更紧密。就力学性能而言,这些材料比较硬而且坚固,但是可延伸(即能够产生大的变形量而不断裂),可以抵抗断裂,这些是它们被广泛用在结构中的原因。金属材料拥有大量不受位置限制的电子,也就是说,这些电子没有被束缚于某些特定的原子内。金属的许多性能都要直接归因于这些电子。例如:金属是电和热的良导体;金属在可见光

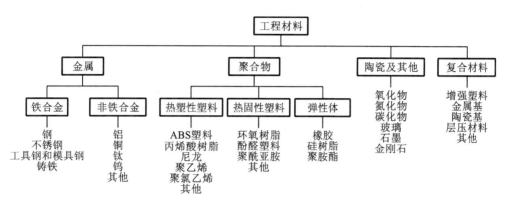

图 3.1 工程材料的分类

下是不透明的;抛光的金属表面具有光泽。另外,一些金属(如 Fe、Co 和 Ni)具有理想的磁性。

虽然人们所用到的元素四分之三都是金属元素,但人们几乎不使用纯金属,这有几个原因。纯金属可能强度太差或者太软,也可能因为稀少而太昂贵,但关键的因素通常是,为获得在工程中所追求的理想性能,需要将金属和其他元素混合起来。因此,这种结合的形式(合金)就得到了很广泛的应用。

根据成分,合金通常被分为两类——铁合金和非铁合金。铁合金的主要组元为铁,包括钢和铸铁。它们是特别重要的工程结构材料,在数量上和在市场价值上所占比例都最大。它们的广泛应用缘于三个因素:(1)铁的化合物大量地存在于地壳中;(2)铁合金可以由相对经济的提取、精炼、合金化和制造技术来生产;(3)铁合金非常通用,因此它们可"定制",以具有多方面的力学和物理性能。

钢是使用最广的合金。铁与不同比例的碳制成合金就形成了低、中、高碳钢。只有碳的质量分数为 0.01%~2.00% 的铁碳合金才被认为是钢。对钢来说,硬度和抗拉强度与碳含量有关,碳含量的增加会导致钢的延展性和韧性变差。然而,热处理如淬火和回火可以显著改变这些性能。最常用的钢的种类有:普通低碳钢、高强度低合金钢、中碳钢、工具钢和不锈钢。不锈钢是有特定规范的钢,含有质量分数超过 10% 的铬。镍和钼也是不锈钢中的典型元素。良好的力学性能加上优秀的耐蚀性使不锈钢能够适应多种用途。铸铁是碳的质量分数超过 2.00% 但低于 6.67% 的铁碳合金。

所有不是以铁为基体的合金都称为非铁合金。重要的非铁合金包括铜、铝、钛、镁及其合金。铜和铜合金拥有理想的综合物理性能,已经有多种应用,比如用于管道系统、器具、导热体、导电体等。铝合金、钛合金和镁合金由于它们有高的比强度而为人们熟知并受到重视,其中镁合金还能用于电磁屏蔽。在某些重要场合,如航空航天工业和某些汽车工程应用中,零部件的体积过大会增加成本投入,这时使用比强度高的材料是很理想的。

第 4 课 机 械 设 计

机械设计是一门通过设计新的或重要的产品来满足人类需求的应用技术科学。它涉及工程技术的各个领域,不仅研究产品的尺寸、形状和详细结构的基本构思,还研究产品在制造、销售和使用等中的多种因素。

机器是机构与其他零件的组合,其用途是转换、传递或利用能量、力或者运动。"机械设计"的内涵比"机器设计"更广,它包括机器设计。对于某些仪器,确定热的流动线路和量这种热力力学及流体力学方面的问题要单独考虑。但是,在机械设计时要考虑运动和结构方面的问题及保存和封装的规定。机械设计在机械工程领域以及其他工程领域都有应用,这些领域都需要诸如开关、凸轮、阀门、容器和搅拌器之类的机械装置。

机械设计包括以下内容:

(1) 对设计过程、设计所需要的公式以及安全因素进行介绍;

(2) 进行材料特性、静态和动态载荷的分析,包括梁、振动和冲击载荷的分析;

(3) 介绍应力的基本原理和进行失效分析;

(4) 介绍静态失效理论和进行静态载荷下机械断裂分析;

(5) 介绍疲劳失效理论,重点是接近高循环的疲劳设计条件下的应力寿命,其通常用在旋转机械的设计中;

(6) 深入探讨机械磨损机理、表面接触应力和表面疲劳现象;

(7) 利用疲劳分析技术研究轴的设计;

(8) 讨论润滑油膜与滚动元件的理论和应用;

(9) 深入介绍直齿圆柱齿轮的动力学、设计和应力分析,并简单介绍有关斜齿轮、锥齿轮和蜗轮的问题;

(10) 讨论弹簧设计,包括螺旋线压缩弹簧、拉伸弹簧和扭转弹簧;

(11) 讨论螺钉、螺杆等紧固件的设计,包括传动螺杆和预加载紧固件的设计;

(12) 介绍盘式和鼓式离合器以及制动器的设计和技术规范。

1. 设计过程

设计是从实际或者假想的需要开始的。现有的设备可能需要在耐用性、效率、重量、速度或成本等方面加以改进;可能需要新的设备来完成以前由人来做的工作,如计算、装配或维修。

设计是一个反复的过程。它始于一种市场需求或一个新的构思,终止于能够满足需要或体现构思的产品的完整说明书。明确需求是首要的,也就是说,要制订一份需求的项目规划书,通常以这样的形式表述:需要一台能实现某功能的设备。规划书的制订者必须客观地强调项目需求(也就是不能暗示如何实现此功能),以避免为成见所囿的狭隘之见。在项目规划书和产品说明书之间有一系列阶段,即初始方案设计、实体设计和细节设计,如图4.1所示。

产品本身就可称为一个技术系统,如图4.2所示,这个技术系统包括部件(或组件)和零件,它们以一定的方式连接在一起以实现需要的功能。就像描述猫时说它由一个头、一个躯干、一条尾巴、四条腿等组成,又将腿分解为股骨、四头肌、爪子、毛皮一样,技术系统由部件组成,各部件又由零件组成。这种分解是分析现有设计的有用方式,但对设计过程本身即新设计的综合分析没有太大的帮助。因此,最好是采用基于系统分析思想的方法,这种方法涉及信息、能量、材料的输入、流动和输出,如图4.3所示。设计将输入转化成输出:电动机将电能转化成机械能,锻压机锻压并使材料成形,防窃报警器收集信息并将其转变成警示音。按照这种方式,可将系统分解成实现特定子功能的连接子系统,如图4.3所示。该图可称为功能结构或功能分解图。这就类似于我们把小猫看做由呼吸系统、心血管系统、神经系统、消化系统等进行相应的连接而形成的。可替代设计用可替代方式连接功能元,联合功能或分解功能。功能结构图给出了系统评估设计的方式。

在设计时,首先要以工作原理为基础来满足功能分解图中的每个子功能,从而完善设计思想。在方案设计阶段(见图4.1),要对所有可供选择的方案进行比较和评定以作出决策。设

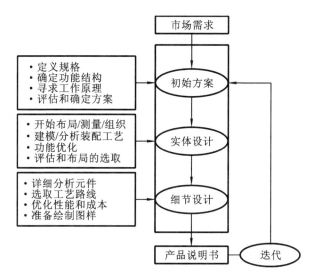

图 4.1　设计流程图

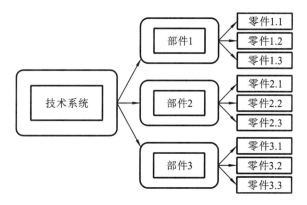

图 4.2　技术系统图

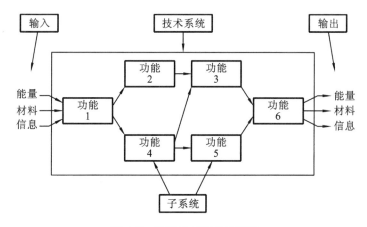

图 4.3　系统分析与能量转换

计者会考虑子功能的备选方案及对这些方案的独立或联合实现方法。下一个阶段,也就是实体设计阶段,对每个有价值的方案,尽力在与实际应用条件相近的情况下去分析其运行状况,确定零件的尺寸,选择合适(在经分析确定的或说明书所要求的一定范围内的压力、温度及环境下会表现良好)的材料,考查其性能和成本。可行设计的产生意味着实体设计方案结束,进入细节设计阶段。在这个阶段,要拟订每一个零件的技术参数,对关键的零件可能要进行精确的力学分析和热分析,在这里,会应用最优化方法来使零件和一系列组件的性能达到最优。最后是选择几何形状和确定材料,进行产品分析和设计评估。详细的产品说明书出来后,该阶段结束。

2. 一些设计规则

为了激发创造性思维,建议设计人员和分析人员使用以下准则。前六条准则尤其适用于分析人员,虽然他可能用到所有这十项规则。

(1) 要有创造性地利用所需要的物理性能并控制不需要的性能。
(2) 认识载荷产生的影响及其意义。
(3) 预先考虑没有作用的载荷。
(4) 发现更有利的加载条件。
(5) 提供最小重量下的合理的强度和刚度。
(6) 用基本方程来计算比例并使尺寸最优化。
(7) 选取不同的材料以获得最好的综合性能。
(8) 在选择备用件和集成零件时要仔细。
(9) 修改功能设计以适应生产过程和降低成本。
(10) 考虑装配中使部件精确定位和互不干扰。

第5课 机械零件(Ⅱ)

1. 传动轴

传动轴是机械动力传动装置中必不可少的元件,主要用于从机器的一端向另一端传递旋转运动和扭矩(见图 5.1)。齿轮、带轮(滑轮)、飞轮、离合器、链轮等零件安装在传动轴上,用来从电动机或发动机向工作机器传递动力。转动力(转矩)通过过盈配合、键、花键和定位销传递到轴上的这些零部件上。传动轴一般用滚动或滑动轴承支承,用推力轴承、键槽、轴肩和定位环等为旋转零部件定位并且承受轴向载荷。根据传动轴的结构和用途不同,常用的轴主要分为以下几种类型。

实心轴　实心轴可以利用直接购买的直径最大可以达到 15 cm 的棒料通过热拔、冷拔或机加工制成,轴的直径可由 6 mm 或更小开始递增。对于大尺寸的轴,常采用专门的轧制加工,而对于尺寸极大的传动轴,常采用锻造方式加工成要求的形状。实心轴往往设计成阶梯轴,轴中间部分直径最大,能承载更大的载荷,轴的末端直径最小,用于安装轴承。另外,阶梯轴的轴肩用于在轴上装配转动零件时对其进行定位。

空心轴　为了减轻轴的重量,有时需要在实心轴内部镗孔或钻孔,或者采用空心管材制造空心轴。空心轴的内部可安装支承件或者其他转动轴。弯曲和扭转强度相等时,空心轴比实心轴的直径更大,但是它的重量较轻。利用铸锭加工大型传动轴时,常在轴中心镗孔用来去除铸造缺陷和检查锻造裂纹等。

轮轴　轮轴是用来支承一个或一组转动轮的轴。轮轴在正常情况下只承受横向载荷,但是有时候也传递扭转载荷。转动轮或齿轮能通过轴承和轴瓦和轮轴配合。轮轴作用的是支承安装于其上的转动轮并使其位置固定。

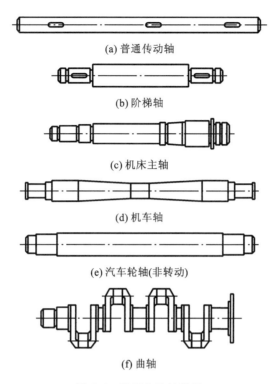

图 5.1 常用传动轴类型

主轴　主轴是一种长度较短、用来安装大型转动零件的转动轴。在所有类型的加工中，主轴都是必不可少的。机床中的主轴用于完成各种加工任务，如磨削、铣削、钻孔、车销、雕刻和蚀除等。高速主轴一般通过法兰与主轴电动机连接，而普通主轴常常通过齿轮或传动带驱动。

传动轴的设计及制造　转动轴的轴向尺寸一般由机构的总体布局确定。传动轴的设计包括确定轴的直径，需要考虑应力集中安全系数和可承受的载荷。普通传动轴一般用中碳钢制造，而对于强度要求高的轴，常使用镍钢、镍铬钢和铬钒钢等合金钢制造。传动轴通常采用热轧成形，通过冷拔、车削或磨削加工至所需要的尺寸。冷轧加工的传动轴强度比热轧加工的高，但会存在较大的残余应力。在对传动轴进行机加工特别是车槽或加工键槽时，残余应力的存在会导致轴发生挠曲变形。大型传动轴一般通过锻造和车削加工而成。

2. 轴类附件

1) 键

键是安装在两个配合部件的键槽中，用于防止两部件发生相对转动或滑动的机械零件(见图 5.2)。通过键可以将转动轴的扭矩传递给转动零件，或者将转动零件的扭矩传递给轴。

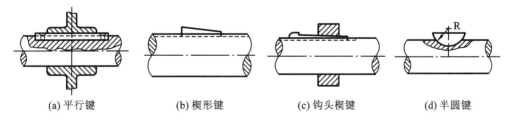

图 5.2　常用键类型

平键　方键和矩形键都是平键,其上下、左右侧面互相平行。平键是目前最标准的键型,应用广泛。一般先将平键安装在传动轴的键槽中,再将轮毂推入,使轮毂键槽的侧面与键配合。

楔键　横键的截面为矩形,但截面积沿键长度方向是变化的。楔形键的最大优点是便于安装和拆卸。另外,由于楔形键沿其长度方向具有一定锥度,常用楔形键对传力零件进行轴向定位。安装楔形键时,先安装轴和轮毂,再将键沿轮毂长度方向推入。

钩头楔键　这种键和楔键十分相似,只是在键尾带有较大的钩头,以便拆装。这种类型的键常被应用于需要经常拆装键的场合。钩头楔键可防止轴向移动,因此这种键能够调整轮毂在轴上的位置并将其固定在最佳位置。

半圆键　半圆键是因它的制造方法而得名的。键截面的形状为半圆形。这种键比平键承受的剪力小。半圆键适用于轻载、易于拆装的场合。另外,半圆键对轮毂和传动轴的对中性比较好,这一特性对高速传动场合,例如涡轮和传动轴的键连接传动尤为重要。

键的类型和尺寸选择通常在轴和轮毂设计完成后进行。一般通过设计分析确定键的长度和材料,其中需要考虑轮毂宽度和键槽至应力集中区的距离这两种因素。键通常由低碳钢和冷拔钢制成,例如 AISI 1020CD,其强度极限为 61 ksi(1 ksi≈6.895 MPa)。键通常由定位螺钉、轴肩、定位环、垫片等固定。

2) 花键

花键通常被描述成由沿轴线方向加工出的多个键组成的键轴(见图 5.3),与其配合的零件(齿轮、带轮、链轮等)的轮毂内开有相应的键槽(见图 5.3)。当所传递的力相对轴的尺寸来说较大时,采用花键轴,如汽车轴、齿轮箱轴等。目前工业上应用的花键类型主要有两种:直齿花键和渐开线花键。后者的强度更高,而且更易于测量,在工作时对中性更好。花键一般通过机加工或轧制而成。由于冷轧温度低,冷轧花键的强度比机加工花键的高。对花键通常采用渗氮处理来获得高表面硬度,以减少磨损。

3) 定位销

为了对安装在轴上的零件进行定位,常在轮毂和轴上钻孔,将定位销插入孔内进行定位。锥形定位销能够防止轴上零件轴向移动和转动(见图 5.4),而圆柱定位销会在轮毂和轴的横截面上承受剪力。如何正确安装圆柱销,实现对轮毂和传动轴的精准定位,并防止销子滑出是比较棘手的问题。锥形销与开口弹簧销部分解决了这一难题。有时采用的销的直径较小,主要是确保销在实际载荷超过设计载荷时能够断裂,以保护重要的机械零件,这种销称为剪切销。

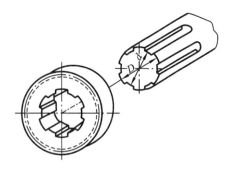

图 5.3　花键

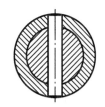

图 5.4　锥形定位销

4) 联轴器

联轴器是指安装在两根传动轴末端用于连接两根传动轴并传递动力的机械装置。目前有两种类型的联轴器:刚性联轴器和挠性联轴器。

刚性联轴器 如图 5.5(a)所示,刚性联轴器用来将两根轴牢固连接,以使它们之间不发生相对移动。刚性联轴器分为套筒联轴器、夹壳联轴器、凸缘联轴器。刚性联轴器只能用来连接在安装和使用过程中对中性都很好的两根转动轴。如果两根被连接的两轴之间有明显的角偏移、径向或轴向偏移,产生的应力将很难预测,这可能导致轴的过早疲劳破坏。采用挠性联轴器能克服这些问题。

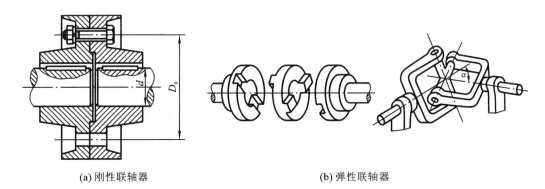

(a) 刚性联轴器　　　　　　　　(b) 弹性联轴器

图 5.5　常用联轴器类型

挠性联轴器 挠性联轴器是为在存在轴向、径向和角度偏移时,平稳地传递转矩而设计的。挠性是指当两轴发生偏移时,联轴器零件能够自由地进行微小移动或者不抵制这种偏移,这样不会产生明显的轴向或弯曲应力。根据所采用弹性元件的不同,挠性联轴器分为无弹性元件的挠性联轴器和有弹性元件的挠性联轴器。前者包含十字滑块联轴器、滑块联轴器、万向十字轴联轴器、齿式联轴器、滚子链联轴器等;有弹性元件的挠性联轴器分为弹性套柱销联轴器、弹性柱销联轴器、梅花形挠性联轴器、轮胎式联轴器、膜片联轴器等。

选择联轴器类型时,需要考虑到以下几种因素:(1)承载转矩大小;(2)单轴或两轴间的轴向运动;(3)轴的对中要求;(4)振动控制情况;(5)轴的挠曲变形情况;(6)设计寿命和成本。

第 6 课　汽车的基础零件

汽车工业是一个快速发展的工业。目前,全世界有数千家工厂在生产汽车,各种汽车产品、机动车辆和发动机有着巨大的市场。现在的汽车平均含有 15,000 多个独立的而又相互联系的零件。有些零件使汽车更舒适或者更美观,但大多数零件的作用都是使汽车能行驶。这些零部件可分为四大类(见图 6.1):发动机、车身、底盘和电气设备。

1. 发动机

在所有汽车零部件中,汽车发动机是最复杂的,而且对汽车的性能有至关重要的影响。因此,发动机通常被称为汽车的"心脏"。内燃机是最常见的,它通过发动机汽缸内液体燃料的燃烧获得动力。内燃机有两种类型:汽油机(也称为火花点火发动机)和柴油机(也称为压燃式发动机)。燃料燃烧产生热量,使气体进入汽缸,导致汽缸内压力增加,将动力通过传动轴传递到变速器。

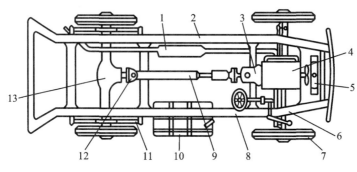

图 6.1 汽车的布局

1—排气装置/消声器；2—车架；3—离合器和变速箱；4—发动机；5—散热器；6—前轴；7—车轮；
8—转向器；9—传动轴；10—燃油箱；11—板簧；12—万向联轴器；13—差速后桥

所有的发动机都有燃料系统、排气系统、冷却系统、润滑系统。汽油发动机也有一个点火系统。点火系统提供电火花，用以点燃汽缸里的空气燃料混合物。打开点火开关时，电流从 12 V 的蓄电池流向点火线圈。线圈电压升高至 20,000 V，从而产生强烈的电火花，点燃发动机燃料。

燃料系统存储液体燃料，并可将燃料送到发动机。燃料存储在油箱，通过油管与油泵相连接，并通过油管从油箱被抽出来，然后经过滤器被送入化油器与空气混合，形成可燃混合气。

发动机的冷却系统必须保持稳定的工作温度，不能太热，也不能太冷。燃料在燃烧过程中会产生大量的热，如果发动机没有自冷功能，它会迅速被烧坏。发动机的主要部件会弯曲，导致水和油的泄漏，同时油会燃烧而不能使用。

有些发动机是风冷式的，但绝大多数发动机都是水冷式的。利用水泵让冷却液在个发动机内循环，流经油缸的发热部位，然后将热的冷却液送到散热器中冷却。

润滑系统在保持发动机平稳运行方面非常重要。润滑系统使用机油作为润滑剂。润滑系统不但可以通过在运动零件上涂油层而减少摩擦，还可以通过在活塞环和汽缸壁之间储存机油，产生密封作用。它还可以通过机油的循环带走沉淀物和冷却发动机。为了保持润滑系统的工作效率，必须定期更换机油滤清器和机油。汽车其他运动部件也需要润滑。

2．车身

汽车车身由钣金外壳制成，车身上制作有车窗、车门、引擎盖、行李箱盖板。它为发动机、乘客、货物提供保护。车身的设计要保证乘客的安全和舒适。通过车身造型，汽车的外形美观、颜色丰富、时尚。它的流线型设计用来减少风的阻力，并防止汽车行驶时摇摆。

3．底盘

底盘是由车辆的主要操作部分组合而形成的。底盘包括动力传动系统、悬挂系统、转向系统、制动系统。

动力传动系统包括离合器、变速器、传动轴、后轴、差速器和驱动车轮装置。

离合器是一摩擦装置，用来连接和切断来自发动机的动力。可手动或自动操作。

变速或齿轮箱的主要功用是提供不同的齿轮传动比（与发动机和驾驶条件相匹配）。齿轮的选择可由司机手动控制或通过液压控制系统自动控制完成。

推进器传动轴（驱动器）的功能是将司机的操作从变速箱传输到后轴差速器总成输入轴。万向联轴器可实现后轴和车轮的上下移动而不影响驾驶。

后轴和差速器经推进器传动轴、后轴到车轮将发动机传递的扭矩改变90°。差速器的另一功能是保证每个驱动轮可以以不同的转数旋转，在转弯时这一功能是必要的，此时外侧轮较内侧轮走过的距离要长。差速器的第三个功能是通过提供另一种传动比来增加扭矩。

悬挂系统的基本功能是吸收不平坦路面引起的震动，否则该震动会被传递给车辆和乘坐人员，因此，悬挂系统有助于在各种路况下保持车辆可控并水平行驶。

转向系统是司机控制方向盘，通过前轮实现转向的。借助转向系统可减少转动方向盘时所需的力量，使车辆更容易操纵。

制动系统有三个主要功能：在必要时能够降低车速；能够让汽车刹车距离尽可能短；能够实现驻车。制动的实现，是固定表面（刹车片）与旋转面（鼓或盘）接触而产生摩擦的结果。每个车轮都有刹车装置，或是盘式的或是鼓式的，由驾驶员用脚踏刹车板，通过液压装置进行控制。

4. 电气设备

电气系统为点火装置、喇叭、灯、热水器和自启动装置供电。电压由充电电路维持。该电路由电池和发电机组成。电池存储电能。发电机将发动机的机械能转变成电能并对电池充电。

第 7 课　制 造 工 艺

制造就是通过对材料、人力、设备和资金的充分利用和管理来制造产品。为了实现成功和经济的制造，从设计开始阶段就必须做好计划，完成材料、工艺、设备的选择和生产工序的安排，这些过程构成了所谓的制造工艺流程。产品的工艺流程包括：产品设计和准备、原材料和已加工材料的运输、毛坯的成形、切削加工、去毛刺、热处理、装配、检验、包装和喷漆等，这些通常是由不同的工厂或者不同车间来完成的。

从原材料到成品的产品工艺：

- 毛坯或者零件的成形——铸造、锻造、冲压、焊接、挤压、烧结、注塑方法等等；
- 切削加工——车削、磨削、特种加工（非传统的加工方法）；
- 材料的性能的处理方法——热处理、电镀和涂覆涂层。
- 装配——根据预定的关系和说明连接零件，包括零件的安装、连接、校正、平衡、检验。

加工的目的是将原材料变成满足产品要求的成品零件。制造工艺关心的是在给定的加工条件下，如何采用既经济又有效的方法和合理的工艺路线来获得理想的零件。

加工金属的五大基本技术包括：钻削、镗削、车削、刨削、铣削和磨削。这五大基本技术各有特色，以满足不同的加工条件（见表7.1，表中 1 in＝25.4 mm）。

表 7.1 韧性材料的基本加工方法比较

加工方法	生产形状	机床	刀具	相对运动 机床	相对运动 刀具	表面粗糙度 $Ra/\mu\text{in}$	公差范围 /in
车削（外部）	旋转曲面（圆柱形面）	车床，镗床	单头	↔ ↕		32～500	±0.001
钻削（内部）	圆柱形面（扩孔）	镗床	单头	↔		16～250	±0.0001 ±0.001
刨削	平面或槽	牛头刨床，刨床	单头	↕ ↔	↔	32～500	±0.001
钻孔	圆柱形面（起始孔直径为 0.010～4 in）	钻床	双边钻孔	↕	固定	125～250	±0.002
端铣面、板	平面，轮廓面，表面和槽	铣床	多头（铣刀齿）			32～500	±0.001
外圆切入磨削	圆柱形面和平面	磨床	多头（研磨轮）			8～125	±0.0001

  钻削是指通过旋转的钻头加工圆孔。而镗削是通过旋转的偏置单头刀具来对已经钻好的孔做进一步的加工。在一些镗床上，刀具是静止的，工件是旋转的，而在有些镗床上正好相反。

  车床是所有机床之父。在车床上，被加工的金属工件是旋转的，切削刀具向着工件运动。

  用龙门刨床刨金属的过程类似于用手刨刨木头。二者的本质区别实际上在于，用龙门刨床刨削时，刨刀在一个位置固定不动，而工件在刨刀下面往复移动。龙门刨床通常是大型设备，有时大到足以加工宽 15～20 ft（1 ft=3.048 m）、长是宽的 2 倍的表面。牛头刨床与龙门刨床的区别在于它的工件是固定的，而刀具往复移动。

  在机械加工中，除了车床之外，铣床使用得最广泛。铣削是利用有多个切削刃的刀具来加工工件的方法。为了满足不同种类工件的加工需求，铣床有许多种铣削形式。有些铣床加工的形状是非常简单的，像用圆锯片铣刀插槽和铣平面。其他铣刀形状较为复杂，可能包含平面和曲面组合，具体取决于刀具切削边缘的形状和刀具的路径。

  磨削是用旋转的砂轮加工工件的方法。这个工艺通常用于零件的最后加工，使其达到设计尺寸，零件应已经热处理而变硬。

第 8 课 公差与互换性

  在制造机器零部件时，质量和精确度是主要考虑的因素。可互换的零件需要高精确度以装配在一起。随着精度的增加和尺寸变动量的减小，制造零件所需的劳动力和机器更趋向成本密集型。任何一个加工人员都应该具有全面的公差知识，知道如何以最低的成本来提高加工零件的质量和可靠性。

为了将设计者的意图向工程图的最终使用者表明,必须在工程图上准确地标注尺寸。设计图给出的零件的尺寸和加工尺寸应该一致,但是不幸的是使零件的尺寸完全符合设计尺寸是不可能的。大多数尺寸都有准确性变化范围和确定尺寸偏差允许极限的手段,以使所加工的零件可以被接受并且有用。零件的尺寸、形状和表面的相互位置必须保证在一定的精度范围内以获得正确和可靠的功能。在常规生产过程中不允许对具有绝对精度的给定几何特征进行修理或测量。因此加工零件的实际表面不同于图样中给出的理想表面。在生产过程中,将实际表面的偏差分成四组来评定、判断和检查许可的不准确度:

- 尺寸偏差;
- 形状偏差;
- 位置偏差;
- 表面粗糙度。

正如上面提到的,生产尺寸绝对准确的机器零件是不可能的。事实上,这也是没有必要和不实用的。零件的实际尺寸在两个极限尺寸之间,许可的偏差确保工程产品的正确的功能,这样就足够了。图样中的尺寸公差决定了给定零件的精度水平。给定的精度与产品的功能性和经济性有关。确定尺寸公差最主要的因素是由尺寸确定的特征的功能。不必要的过于严格的公差将会导致高的加工成本,高的加工成本缘于高成本的加工方法和较高的废品率。

1. 尺寸和公差

在图样上标注尺寸时,尺寸线上标注的数字仅仅表示近似的尺寸,而不表示任何精度,除非设计人员加以说明。这个数字称为基本尺寸。设计人员主要强调零件的基本尺寸以使其达到要求。实际上,由于表面的不规则性和表面粗糙度,零件不可能恰好加工到基本尺寸。必须允许尺寸有些变动以确保加工的可能性。但是,允许的变动量不能太大,否则装配零件的性能就会被破坏。单个零件尺寸允许的变动量就是公差。

为了保证零件的安装和互换性,控制尺寸是必要的。公差标注在影响配合的间隙和过盈的主要尺寸上。标注公差的一种方法是在公称尺寸后面紧跟着允许的变动量。例如 40.000 ± 0.003 mm,表示加工的尺寸在 39.997 mm 和 40.003 mm 之间。如果变动量在公称尺寸的两边,公差就叫做双边公差。如果一边公差是 0,这种公差就叫做单边公差。

大部分部门有一些通用公差,当图样上不标注详细的尺寸时就应用这样的公差。对于加工尺寸,通用的公差是 ±0.5 mm,因此如果标注的尺寸是 15.0 mm,加工尺寸就在 14.5 mm 和 15.5 mm 之间。其他的通用公差应用在圆角、钻削孔和冲压孔,铸造、锻造焊接件和倒角上等特征上。

2. 孔和轴的标准配合

一项一般性的工程工作是确定一个圆柱形零件的公差,比如固定在一个相配合的圆柱形零件或孔内的,或者是在其内部旋转的轴的公差。配合的松紧取决于应用。例如,一个定位在轴上的齿轮就需要"紧"配合,这里轴的直径应比齿轮毂的内径稍微大一些以便能传递所需的扭矩。又如,使滑动轴承的直径大于轴的直径以使其可以旋转。考虑到从经济角度出发,不可能把部件尺寸制造精确,就必须确定轴和孔尺寸大小变化量。但是,变化的范围不应太大,以免造成装配不便。为了使被指定的公差尺寸不至于过多,公差带国家和国际标准已经制定,实例列在表 8.1 中,比如 H11/c11。为了将这些信息转化为实际的尺寸,相应确定了不同规格尺寸的公差表。为了使用这一资料,图 8.1 给出了传统公差中术语的定义。通常使用的是基孔制,因为这会使企业内部所需的钻头、铰刀、拉刀和计量工具种类减少。

表 8.1　公差带及其典型应用的实例

分类	说明	特性	ISO 符号	装配	应用
间隙配合	松的转动配合	适用于宽的普通公差	H11/c11	间隙较大	气门座内的内燃机排气门
	自由转动配合	适用于温度变化大、转速高或轴颈压力大的场合	H9/d9	间隙较大	多轴承轴、液压缸内的活塞、活动杠杆、滚动轴承
	紧的转动配合	适用于在精密机械上转动的场合和在中等速度和轴颈压力下的精确定位	H8/f7	有间隙	机床主轴承、曲柄轴和连杆轴承、轴套、离合器套、导向块
	滑动配合	当部件不用自由转动但必须精确移动、转动和定位时	H7/g6	没有显著间隙的压入配合	推进齿轮和离合器、连杆轴承、指示器活塞
	定位间隙配合	为固定件的定位提供紧密配合，但可以自由装配	H7/h6	用手压入，带润滑	齿轮、尾架套筒、调整环、活塞螺栓
过渡配合	定位过渡配合	适用于精确定位（位于间隙配合与过盈配合之间）	H7/k6	容易用锤子敲	滑轮、离合器、齿轮、飞轮
		适用于更精确定位	H7/n6	需要压力	电动机轴衔铁、轮子上带齿的轴环
过盈配合	定位过盈配合	适用于要求有刚性和定位精确对中的部件	H7/p6	需要压力	开式滑动轴承
	中等驱动配合	适用于普通钢机件或小截面上的热压配合	H7/s6	需要压力或温度差	离合器从动盘毂，轴承滑块、轮子和连杆内的轴承轴瓦，灰铸铁轮轴上的青铜凸缘

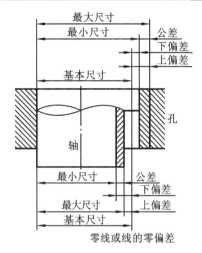

尺寸：以专用单位表示尺寸数字值的数。
实际尺寸：通过测量得到的部件尺寸。
极限尺寸：特征允许的最大和最小尺寸。
上极限尺寸：两个尺寸极限中较大的。
下极限尺寸：两个尺寸极限中较小的。
基本尺寸：用以确定尺寸极限的参考尺寸。
实际偏差：一个尺寸与其对应的基本尺寸的代数差。
公差：上极限尺寸与下极限尺寸的差值。
轴：指定一个部件所有外部特征的术语。
孔：指定一个部件所有内部特征的术语。

图 8.1　传统公差中术语的定义

第 9 课 数 控 技 术

今天,传统的机床普遍已被数控(CNC)机床取代。数控机床仍旧保持传统机床的基本功能,但机床的运动是通过电子控制而不是通过人手控制的。数控机床能以较小的误差重复生产同样的零件,而且可以连续运行,不会疲惫。这些都是数控机床与传统机床相比较所具有的明显优势,传统机床在做任何事情时都需要大量的人机交互过程。

这并不是说,传统的机床过时了。它们仍然在工具和夹具加工、维护和修理,以及小批量生产中被广泛使用。不过,目前许多大中批量的生产工作是由数控机床完成的。此外,计算机数控加工已在小批量的车间得到普及,在这些地方生产十几个到几百个零件这样的规模是常见的。

数控机床有较高的生产率,同时购买、配置和维护费用也高。然而,如果使用时管理得当,生产率的优势很容易就可以抵消这一成本。以选择传统方式还是计算机数控方式生产零部件,主要由配置成本和生产批量决定。对传统机床进行加工操作的配置成本是相当低的,而数控机床的配置成本可能相当高。总的来说,由于生产量增加,生产率提高,配置成本能分散到众多部件上,数控加工成为显而易见的选择。

数控机床设备复杂,对其更详细的研究是一个专门的课题。然而,在一般情况下,任何数控机床都由以下几个单元组成:计算机、控制系统、驱动电动机、换刀装置。

根据数控机床结构(见图 9.1),数控机床按以下方式工作。

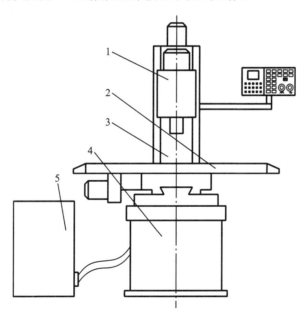

图 9.1 数控机床

1—主轴承;2—工作台;3—立柱;4—底座;5—机器的电气控制系统

(1) 数控机床读取准备程序,并转换成机器语言。机器语言是在计算机上使用的二进制编程语言,该语言不应用在数控机床上。

(2) 当操纵者开始执行循环时,计算机将二进制代码转换成电子脉冲,自动发送到机床的动力单元。控制单元比较发送和接收的脉冲数。

(3) 当电动机收到每个脉冲时,它们会自动将脉冲转换成主轴和丝杠的旋转,使轴转动、移动或使工作台移动。这样,工作台上的零件或车床刀架上的刀具就被运送到程序指定的位置。

还有一些相关的概念,如数控(NC)和加工中心。

数控(NC)是一种通过数字、字母和其他符号实现对工艺设备的控制的可编程自动化控制技术。为实现特定的任务(部分或全部任务),这些数字、字母和符号以一种适当的格式被编制为程序指令。当任务变动时,程序指令将发生改变。

数控系统由下列部件组成:数据输入装置、带控制单元的纸带阅读机、反馈装置和切削机床或其他形式的数控设备。

数据输入装置也称"人机联系装置",可用人工或全自动方法向机床提供数据。人工方法作为输入数据唯一方法时,只限于少量输入。人工输入需要操作者控制每个操作动作,这是一个既慢又单调的过程,除了简单加工场合或特殊情况下,已很少使用。

虽然纸带上的数据是自动给进的,但实际编程却是手工完成的,在编码纸带做好前,编程者经常要和一个设计人员或工艺工程师一起工作,选择合适的数控机床,确定加工材料,计算切削速度和进给速度,确定所需刀具类型。编程者要仔细阅读零件图上的尺寸,定下合适的程序开始的零参考点,然后写出程序清单,给出描述加工顺序的编码数控指令,使机床按顺序加工工件到图样要求。

控制单元接收和存储所有数据,直到积累出一个完整的信息单元为止,然后解释编码指令和引导机床实现所需的运动。

装在纸带阅读机里的硅光二极管正对着控制单元,检测穿过移动纸带上的孔漏过的光线。光束被转变成电能,并通过放大以进一步加强信号,然后信号被送入控制单元里的寄存器,由控制单元将动作信号传送给机床驱动装置。

加工中心可以被定义为这样的一种设备:
(1) 在一个单一利用多轴的设备中进行多项操作和展开多个流程;
(2) 通常有一个自动换刀机制;
(3) 运动的动作是可编程的;
(4) 由伺服电动机控制进刀机构的进给;
(5) 位置反馈由检测机构提供给控制系统。

加工中心主要有两种类型:卧式加工中心和立式加工中心。

卧式加工中心 (1) 立柱移动式加工中心通常采用一个或两个工作台,在工作台上可以安装工件。使用这种类型的加工中心时,在加工工件的同时,操作者可以将新的工件安装在其他工作台上。

(2) 固定式加工中心配有一个托盘式往复移动送件装置。托盘是一个可移动的平台,用来安装工件。工件加工完后,工件和托盘被移动到往复移动送件装置中轮转,将新的托盘和新工件送至加工位置。

立式加工中心 立式加工中心是床鞍式机床,带有滑动导轨,床鞍利用主轴头垂直滑动,而不是依靠主轴运动。

第 10 课 材 料 成 形

材料成形是将工件的原有形状变成另外一种形状而不改变工件材料的质量和化学成分的一系列工艺方法的总称。在此工艺过程中,材料产生的应力大于材料的屈服强度,但小于材料的断裂强度。应力的类型可以是拉应力、压应力、弯曲应力或切应力,或者是这些类型应力的组合。这是一种很经济的方法,因为可以获得所需的形状、尺寸和粗糙度而无须使材料有任何显著的损失。此外,通过应变硬化,一部分输入的能量在提高产品强度上得到了很好的利用。

1. 冷成形和热成形

成形工艺可以分为两个大类:冷成形和热成形。如果加工温度高于材料的再结晶温度,那么这一过程就叫热成形,否则就称为冷成形。

许多冷加工工艺方法可以用于材料的成形并使金属得以强化。例如,轧制用于生产金属板材及薄板等。锻压是将工件材料置于模具型腔中,以获得形状相对复杂的零件,如汽车的曲轴、连杆等。拉深工艺用于制造铝合金饮料罐等。挤压时,材料被推入模具中挤压以获得截面形状均匀的产品,如棒类、管类产品及门窗用的铝合金镶边等。拉伸、弯曲及其他工艺方法都可以用于材料的成形。所以,冷加工是一种成形金属同时又使金属强化的好方法。用冷加工成形可以获得尺寸公差小且表面质量好的零件。

热变形加工非常适用于较大零件的成形,因为金属在高温时屈服强度低而塑性很好。另外,具有密排六方晶体结构的金属,如镁等,在高温时,滑移系会增多。高塑性使热成形有比冷成形更大的变形量。如一块厚板可以经过一系列的连续热加工变成薄板。热加工的优点是通过加工能消除材料中的缺陷:可消除原材料中的一些缺陷或将其影响减至最小;气孔能被压合或熔合;材料中的成分偏析会减少;通过再结晶可以很好地细化及控制金属的显微组织。因此,金属的力学性能和物理性能都能得到明显的改善。

典型的成形方法有轧制、锻造、拉深、拉延等。为了更好地理解各种成形操作的机械学原理,我们将简单讨论一下每种方法。

2. 轧制

轧制是对固态金属材料进行初步机械加工,使其产生塑性变形但质量守恒的一种工艺方法,广泛应用在板材、薄板和结构桁条等的制造中。图 10.1 所示为板材或薄板的轧制。铸造生产出钢锭,经过几个轧制阶段,其厚度不断变薄。轧制通常在钢锭温度较高时进行。由于工件的宽度保持不变,工件将随着厚度的减小而变长。在热轧阶段之后,最终阶段是进行冷却,以提高表面质量、精度,并提高强度。在轧制工艺中,根据需要,轧辊的外形被设计生产成所期望的几何形状。

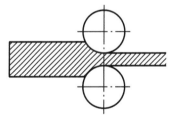

图 10.1 轧制工艺

3. 锻造

锻造是最重要的制造方法之一。锻造同挤压类似,属于塑性变形,但又不同于挤压,它可以用来加工复杂的三维零件。锻造可用来加工从非常微小到重达数吨的各种形状和规格的零件。锻造可以分为两大类:开式模锻和闭式模锻。

开式模锻 图 10.2(a)为开式模锻的工艺过程示意图。在开式模锻中,工件至少有一个表面发生的是自由变形,因此,采用开式模锻加工出的零件在精确度和尺寸公差方面比采用闭式模锻加工出的都要差。但是,它的工具简单,造价相对较低,而且,设计制造比较容易。

闭式模锻 在闭式模锻工艺中,工件被完全限制在模具当中,并且无飞边产生,材料的利用率非常高。在锻造前后工件的体积没有变化,因此,原始坯料的体积控制变得至关重要。多余的材料会产生很大的压力,这容易导致模具失效。

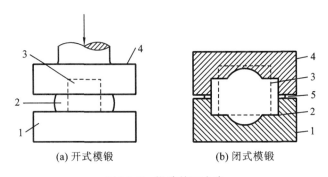

图 10.2 锻造的两大类

1—模具 1;2—最终形状;3—初始形状;4—模具 2;5—飞边

4. 拉深

在拉深中,杯形产品是借助于一个凸模和一个凹模由一块金属薄板获得的。图 10.3 是这一工序的示意图。金属薄板被放在模具上,利用一个压边圈来避免产生缺陷。

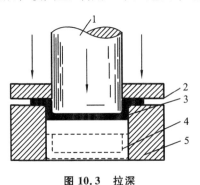

图 10.3 拉深

1—冲头;2—压边圈;3—工件;4—最终产品;5—模具

许多零件都可由这种方法制造,成功的操作需要仔细地控制诸如毛坯夹持器压力、润滑、间隙、材料特性和模具的几何形状等要素。综合考虑很多因素,毛坯直径与凸模直径的最大比值为 1.6～2.3。

人们对该方法进行了广泛研究,结果显示材料的两种重要拉深性能是金属应变硬化指数和压变比(各向异性比)。当材料进行拉深时,前一个性能占主导地位;而纯径向拉延时,后者起主要作用。压变比被定义为一块金属板条在宽度方向上的真实应力与其在厚度方向上的真实应力比值。这个比值越大,金属在抵制厚度变薄并在宽度方向上改变的能力就越强。

薄板平面的各向异性会导致褶皱,即在拉深的板上出现波浪状的边缘。凸模与凹模间隙是这一工艺中的另一个要素,间隙通常被调整为不大于薄板厚度的 1.4 倍。间隙过大,制造的板的厚度就会沿着竖直方向增加,而间隙正确时通过压扁制造的板的厚度就均匀。同样,如果毛坯夹持器的压力过小,边缘就会起皱;如果过大,板就会因增大了的边缘摩擦阻力而被冲掉。

对于相对厚一些的薄板,有可能通过专门的模具设计拉深部件而不需毛坯夹持器。

5. 弯曲

顾名思义,弯曲是对金属薄板进行塑性弯曲而获得所需形状的工艺。弯曲工艺由一套设计好的凸模和凹模来完成。典型的工艺原理如图10.4所示。

弯曲成形可以在剪切工艺设备——偏心曲柄滑块压力机上完成。在弯曲工艺中,材料内部既有压应力,又有拉应力,其值低于材料的极限应力,同时材料的厚度无明显变化。在折弯机上,薄板可以进行简单的直线弯曲。还有一些弯曲操作,如卷曲、Z形弯曲和折边等也与此类似,尽管其工艺略有改进。

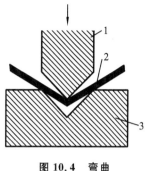

图10.4 弯曲

1—冲头;2—工件;3—模具

第11课 柔性制造

柔性是指以小批量生产一定数量、不同类型的产品,而生产力则意味着高速生产,即自动化大批量、大规模生产。在一般情况下,提高生产率同时保持柔性已经成为大多数行业的主要目标之一。柔性制造系统(FMS)的出现正是为了顺应这一方向。换句话说,FMS试图同时实现柔性和生产力来满足当今竞争激烈的市场中不断变化的客户需求。在过去的几十年中,柔性制造系统经历了快速的演变和发展,并且在一些制造企业中已经实现了许多行之有效的系统。FMS定义为一种由计算机控制的工作站和物料处理系统组成,用来进行多品种、中小批量零件生产以满足客户需要的制造系统。如图11.1所示,FMS基本上是加工设备、物料储运设备和计算机控制系统软件的组合。现代工业中存在不同种类的FMS,其中最具代表性的是柔性制造单元、柔性制造系统和柔性制造线。

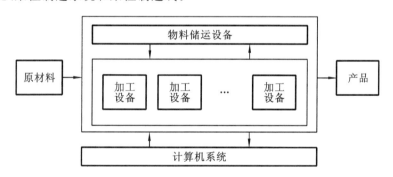

图11.1 FMS基本结构示意图

柔性制造系统的关键组件包括:

(1) 按照零件加工顺序安排的加工机床,这些机床通常是执行某些零件的加工任务的CNC(计算机数控)机床。在柔性制造系统中,还可能包括其他类型的自动化工作站,例如检测设备。

(2) 自动化物料运储系统,例如一个高架自动装卸系统。为完成零部件传输任务,利用柔性制造系统将该装卸系统和输送机连接在一起。物料运储系统也与所有的机床实现了电气连接以进行电子信息交换,从而控制零件的装卸。

(3) 中央计算机系统,负责每台机器与物料运储系统的内部连接以及信息交换,以及协调自动化物料运储系统中的机械爪的运动以及时装卸工件。在柔性制造系统中，零件的加工和处理程序预先加载到每台机床上，用来生产不同类型的零件。

传统的柔性制造系统是由一系列机床所组成的，而这些机床之间又通过零件传输系统相互连接。在制造业发展的初期，柔性制造系统通常应用在加工车间的设计和布置方面，在这些加工车间可以生产小批量、不同类型的零件。早期柔性制造系统的主要优点是能够利用同一组机器设备加工多种类型的零件。对于为需求产品数量不是很大、但是种类较多的客户提供服务的供应商，这种系统是理想的选择。

随着计算机科技的发展及其在工业中的应用，自20世纪70年代至今，诸如可编程逻辑控制器（PLC）、计算机数控（CNC）这样的先进过程控制技术相继被人们开发出来。以更高的生产率和效益实现批量化的生产，以及不需要大的换装成本也能够生产不同零件，这些需求直接促进了更高效的柔性制造系统的设计和产生，这种系统具有自动控制和物料运储能力。世界范围内的汽车工业以及其他行业的主要制造领域中都应用了这种现代化的柔性制造系统。这种系统是为生产大中批量、多批次不同尺寸的产品而设计的，具有柔性，可以从一种类型产品的生产快速地转换到另一种类型产品的生产。

由于柔性制造系统具有成本效益和运营优势，美国汽车行业中包括原始的汽车制造商和零件供应商在内的一些企业已经接受柔性制造的概念，并且已经在其国内外企业中成功地使用了柔性制造系统。由于汽车行业的竞争激烈，美国以及世界上的其他国家拥有汽车的居民数量又庞大，汽车工业中的生产系统必须在实现高效率、大批量生产的同时，能够快速地转换产品类型以适应市场需求的变化而不会产生太大的换装成本。在汽车行业中，设计和管理理念都具有优势的柔性制造系统和专用生产线相比是一种更好的解决方案。成功实现柔性制造策略的公司，能够从大批量（规模经济）、多样化（范围经济）生产中实现长远利益。能在生产批量和产品种类两方面都做好的公司，在业务持续增长和长远发展方面，都会有一个更好的前景。

柔性制造系统的优势包括：
- 能速度更快、成本更低地从一种产品的生产转换到另一种，有助于提高资本的利用率；
- 能减少操作工人数量，可以降低直接劳动力成本；
- 由于采用自动控制，使得库存量减少；
- 由于使用的工人数量不变但生产能力有较大提高，单位产品的成本降低。
- 在直接劳动力、故障率的降低、返工、修理和废品等方面都可节省成本。

制造企业正不可避免地面临着大量的不确定性因素和生产过程中的不断变化。企业需要处理这些问题或者在问题出现之后作出反应。不确定和突发事件可能改变系统的状态，并且影响系统的性能。生产计划的制订作为制造业系统中的一个关键的环节，目的在于提高生产力并降低运行成本，然而生产计划却经常被计划执行过程中的一些未知因素打乱。机床可能在运行过程中出现重大的故障而停止运行，例如机床损坏、刀具磨损或者刀具需重新调度。如果这些因素可能导致系统性能的大幅下降，系统需要作出反应并且实时修改当前的计划，以减轻这种影响。FMS主要缺点是：
- 对零件或零件组变换的适应能力有限；
- 需要大量的前期规划活动；
- 成本相对较高；
- 存在精确的组件定位和精确定时的技术问题；
- 制造系统结构复杂。

第12课 机电一体化

近年来,市场对多元化产品的需求不断增长,直接推动了机械工程学科的发展。其中最常见的例子是将机械设备与电子电路有机地结合为一体,以发挥两者的基本功能。最简单的机械装置,如齿轮、滑轮、弹簧以及带轮等构成了机械工业的基础。另一方面,电子技术是20世纪的产物,所有电子技术的产品都是在过去的100多年内被创造出来的。机械装置在工业中起着重要作用。在很多工业过程中使用电子控制系统的同时,仍需要机械装置传送动力来进行工作。机电一体化是指将机械和电子工程学科整合在一起,以实现发展智能控制的高效系统的目标。国际机器理论与机构学联合会将机电一体化定义为在产品设计和制造过程中,机械工程、电子工程、电子控制以及与产品设计和制造过程有关的计算机软件的有机结合,如图12.1所示。被创造出来的机电一体化比其字面上的意义更为重要,它正在改变我们设计和生产新一代高科技产品的方式。

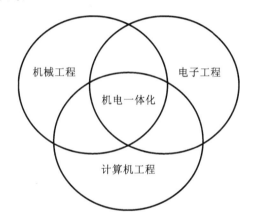

图12.1 机电一体化由机械、电子和计算机工程融合示意图

机电一体化是跨学科的技术学科,建立在经典的机械、电气和电子工程的基础上。这三门学科之间不仅彼此结合紧密,而且还与计算机科学和软件工程相关联。机电一体化的研究关键在于通过智能控制原理对系统各机构部件进行组合研发。因此,一个系统由机械部分和电子部分组成,具体包括用来记录信息的传感器、用来分析处理信息的微控制器、根据这些信息作出反应的程序集合,这样就构成一个完整的机电一体化系统,如图12.2所示。

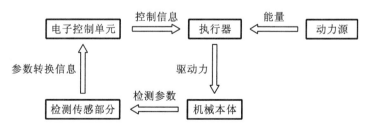

图12.2 机电一体化系统的基本结构

现代工业系统和零部件通常采用包括多种传感器、执行器和控制器,并将其集成到复杂机构当中,这些机构则是各种工程技术有机结合。为了设备和系统的设计及维护,全球性企业通常会采用熟悉机电一体化系统技术的工程团队。其中必要的关键专业技能包括机械、电气、计

算机、工业工程,以及控制系统、计算机仿真、机器人技术和人力资源管理技术,例如,使用不同类型的传送机构和机器人来运送物资、组装零部件,并且在生产线上传送成品就需要以上技能。

机电一体化系统设计的基本思想要求在一个机电一体化系统的设计过程中应包括交叉学科部分和系统的集成。这种集成/并行的系统设计方法的目的在于提高系统性能,而单独利用各传统学科的设计方法难以达到同样的效果,这种独立的设计方法明显忽视了系统各部分之间的相互影响。机电一体化系统设计的一个主要问题是软、硬件部分的集成以及机电一体化应用产品设计中所涉及的各种技术学科之间的相互交叉及有机整合。随着机电一体化系统体积的增大和复杂性的提高,也就是说,大规模的机电系统,其系统设计超出了硬件和算法的范畴。为了确保经济性、可靠性、可维护性、适应性和由大规模应用带来的增长潜力,需要对机电一体化系统软、硬件设计过程进行根本性的改进。在下一代智能系统的开发过程中机电一体化技术面临的重大挑战之一,就在于机电系统的多学科综合的特性对硬件和软件的集成提出了进一步的要求。这些大型系统的设计和集成需要在比以往更高的层次上展开。

第 13 课　液压与气压系统

1. 液压系统

液压系统是以液压油为工作介质来推动液压机械的一种驱动或传动系统。液压系统由油箱、液压泵、控制阀和液压马达等组成,通过管路和接头连接形成一个液压回路(见图 13.1)。

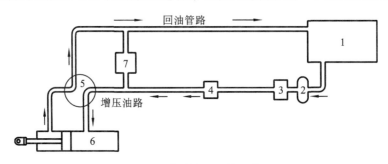

图 13.1　典型液压系统示意图
1—油箱;2—油泵;3—过滤装置;4—压力开关;5—换向阀;6—执行机构;7—减压阀

液压泵 3 在工作时由油箱 1 中抽吸液压油,然后推动液压油经过换向阀 5 进入整个回路中。由于液体的不可压缩性,液压油产生压力,执行机构 6 随即开始向左移动,同时位于执行机构 6 左侧的液压油经过换向阀 5 流回油箱 1。液压油进入液压泵 3 之前先经过过滤器 2,以确保没有杂质进入液压回路中。执行机构 6 推动物体前行。液压系统设定的最大压力只能与推动物体所需压力相等。当液压回路的压力超过设定值时,压力控制开关 4 开启,溢流阀 7 工作,多余的液压油通过旁路流回油箱中,从而防止压力过大而损坏液压系统。

1) 液压泵

液压泵的作用是为液压油增压以确保液压系统正常工作。每个液压系统可以利用一个或多个液压泵为液压油增压,具有一定压力的液压油进而被输入液压系统工作区而做功,这样就可以利用液压油推动液压缸中的活塞或驱动液压马达等。在液压机械中,常用的液压泵类型分为齿轮泵、叶片泵和柱塞泵等(见图 13.2)。

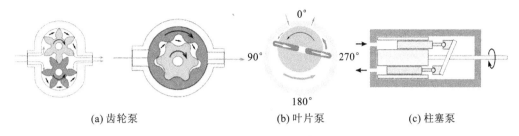

(a) 齿轮泵　　　　　　(b) 叶片泵　　　　　(c) 柱塞泵

图 13.2　常用液压泵类型

齿轮泵是容积泵，通过齿轮的啮合实现吸油和压油。齿轮泵是液压传动系统中最常用的液压泵。齿轮泵有两种类型，即外啮合齿轮泵和内啮合齿轮泵。外啮合齿轮泵采用两个外部啮合的直齿圆柱齿轮，而内啮合齿轮泵由一个外直齿圆柱齿轮和一个内直齿圆柱齿轮组成。齿轮泵为正压容积泵，即每转一周泵出的液体容积为定值。某些齿轮泵既可以作为液压马达也可以作为液压泵使用。

叶片泵亦为正压容积泵，叶片安装在转子上，转子在泵壳内腔旋转。某些结构的叶片泵，泵旋转时叶片的长度是变化的，同时(或)通过预紧力使叶片与泵壳内腔壁接触。在泵的吸入端，叶片与泵内腔所形成的容积逐渐增加，液体在入口压力的推动下被泵入其中。叶片泵的入口压力源自泵系统的运转，常为大气压力。在泵出端，叶片与泵内腔所形成的容积逐渐减小，迫使液压油从泵中流出。

柱塞式液压泵比齿轮泵和叶片泵造价高，适用于高压工作环境和泵送特性复杂的流体，性能可靠、使用寿命长。轴向柱塞泵中有多个(通常是奇数个)柱塞，柱塞按圆周方式排列在缸体、转子或者柱塞套内。当柱塞向阀板移动时，推动液压油由阀板上的排油口排出。径向柱塞泵通常用于压力高、流量小的场合。在径向柱塞泵中，柱塞沿径向、对称安装在驱动轴周围，并沿轴的径向运动。吸油时活塞由内止点开始移动，旋转 180°后，柱塞泵内充满液压油，此时柱塞处于外止点，将泵入的液体排出。

2) 控制阀

液压阀用来调节液压管路中液体流向、工作压力、流量，进而调节流速。有四种典型的控制阀，使用时根据实际问题进行选择。

方向控制阀　方向控制阀控制液压油的流向，进而控制工作元件的运动方向及位置。控制阀通常由滑阀柱塞和缸体构成，柱塞由机械力或电力控制。通过移动柱塞导通或限制流体流动，以此控制流体。

压力控制阀　压力控制阀用于保持液压回路中各处压力的稳定。压力控制阀通过两种方式调控回路中的压力：将高压流体导入低压区域；限制流体进入其他区域。改变流体流向的控制阀类型有：安全阀、溢流阀、平衡阀、顺序阀和卸荷阀。减压阀可限制流体流向回路中的其他的区域。溢流阀、顺序阀、卸荷阀和平衡阀属于常闭阀，根据预设功能要求，这些阀在工作时处于全开或部分开启状态；减压阀一般为常开阀，用来限制和阻断液压油进入下游工作区。

流量控制阀　流量控制阀一般与压力控制阀联合使用来控制液压回路中流体的速率。控制或调节动力元件的工作速度并保持流体速度恒定，进行分流，这一般是在流量控制阀与压力控制阀的联合作用下实现的。

单向阀　单向阀又称止回阀，是使液体只能沿一个方向流动的机械装置。单向阀有多种类型，可以广泛应用于不同场合。虽然单向阀有多种规格，而且价格各异，但总的来说都体积小、结构简单,价格也可能较便宜。单向阀可以自动控制,很多时候不需要人工操作或其他外部控制措施,因此,单向阀一般没有阀杆或手柄。

3) 液压系统的优点

可以实现无级调速和换向 液压系统的执行机构(直线运动或旋转运动)可以通过流量控制阀调节液压泵的排量,以达到调控速度的目的。液压装置可以在运行的过程中进行调速。另外,液压系统的执行机构可以在满载时实现快速换向,并且不会受到损害。

承载能力高、体积小 由于液压油几乎不可压缩,弹性变形非常小,因此液压回路中的液压油可提供相当大的动力,移动相当高的工作载荷。另外,由于液压元件能够耐受高速和高压,利用体积非常小和重量非常轻的液压元件也能输出较大的动力。

具有过载保护功能 液压回路中的溢流阀能够防止系统因过载而受到损害。当系统中的压力超过溢流阀设定的压力值时,液压泵会直接将液压油输到油箱中,使系统输出的转矩或动力保持恒定。溢流阀也可以为工件夹紧机构等机械设置提供指定的工作转矩或动力。

2. 气压系统

气压传动是研究利用压缩气体推动机械运动的一种技术。气动回路由以下几部分构成:空压机、气缸、储气罐、控制阀、气动软管等。图 13.3 所示为一个带有节气调节装置的气动回路。图中所示的二位五通换向滑阀带有双压力进气管,在节省罐中压缩气体的同时可以保证回路正常的工作循环。压力控制阀所设定的回程压力必须达到缸体中活塞杆一端运动到终点的最低压力,以确保工作的连续性。当回路压力降至 20 psi(磅/平方英寸,1 psi = 6.895 kPa),低于工作要求的压力时,压力控制阀会快速启动。改变五通阀流向,气缸开始吸气。由于处于活塞杆一端的低压空气受到压缩,会阻碍活塞柱头一端的高压空气进入,因此吸气过程中会存在短暂的冲击。为了有效控制气压回路的工作周期,可通过调节与活塞杆一端相连的出口流量调节阀来控制活塞移动的速度。当改变五通滑阀的流向,使气缸活塞返程时,须将与活塞环一端相连的排气流量控制阀的流速调整到最大值,因为返程动力的供给是有限的。

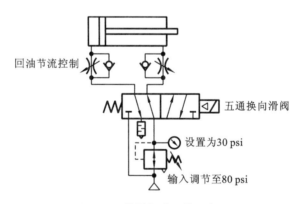

图 13.3 典型气动系统示意图

气压技术被广泛应用在各种行业,特别是工业自动化领域中以驱动自动化机械。气动系统的优点包括以下几点。

1) 结构简单、容易操作

气压元件的设计相对比较简单,因此它们适用于简单的自动化控制系统。另外,气压系统的直线运动和往复运动容易调节,而且限制因素少。气压回路中的空气压力和体积可以容易地通过压力调节阀控制。由于气压元件造价低廉,气压系统的成本非常低。另外,气压系统运行可靠,其维修成本显著低于其他系统。

2) 可靠性和安全性高

许多企业的生产线上装有由移动式压缩机提供压缩空气的气压系统。而且,气压系统中压缩空气的供给不受距离的限制,可以通过管道输送空气。由于气压系统在易燃环境中使用不会引起火灾或爆炸,因此气压系统比电动系统更为安全。另外,当气压系统发生过载时,仅会出现轻微的滑动或运行中断。

3) 适用于恶劣的工作环境

与其他系统相比,气压系统中的压缩空气不受高温、粉尘、腐蚀等影响。气压系统的运行不会产生污染,因此,气压系统可以用在清洁度要求较高的场合,例如集成电路的生产线中。

液压传动和气压传动都是流体动力技术的应用。气压系统的工作介质是可压缩的气体,如空气或其他适宜的纯净气体,而液压系统的工作介质是不可压缩的液体,如液压油。工业上应用的大部分气压系统的压力在 80~100 psi(550~690 kPa)之间,而液压系统的压力在 1000~5000 psi(6.9~34 MPa)之间,在某些特殊场合,其压力会超过 10000 psi(69 MPa)。

第 14 课 现代设计理论与方法

现代设计理论与方法是指新兴的,传统的重点关注强度和刚度的与那些设计理论和方法不同的设计理论和方法。它们的特点有:用计算机进行优化设计、计算机辅助设计、可靠性计算等。因此,现代设计理论与方法覆盖了多种主题,包括所有非传统的新兴设计理论和方法。现代设计理论与方法一般包括以下内容。

1. 优化设计

优化设计是从多种方案中选择最佳解决方案的设计方法。它以最优化理论和计算机计算为基础,根据实际问题确定设计变量,建立目标函数,在满足给定的各种约束条件下,寻求一个最优的设计方案。第二次世界大战期间,在军事上优化技术最先得到应用。机械优化设计起始于 20 世纪 60 年代,随后广泛应用于生产。1967 年,R. L. Fox 和 K. D. Willmert 发表了他们运用优化设计技术研究曲线发生机构的成果。此后,C. S. Beightler 等人用几何规划解决了液体动压轴承的优化设计问题。机械优化问题通常属于非线性规划范围,随着数学理论和计算机技术的发展,优化设计已广泛应用于工程,并成为一个独立的工程学科。

优化设计问题的数学模型由以下三要素组成:设计变量、目标函数和约束条件。优化设计问题可以这样描述:寻找一组设计变量,能满足所有的约束条件和优化目标函数。一个常见约束优化问题的数学模型可记作

$$\begin{cases} \min f(x), & x \in D \subset \mathbf{R}^n \\ \text{s.t.} \quad h_v(x) = 0, & v = 1, 2, \cdots, p \\ \quad g_u(x) \leqslant 0, & u = 1, 2, \cdots, m \end{cases}$$

2. 摩擦学设计

摩擦学设计在节能减排和环保方面起着重要作用。它是机械产品设计的一个重要部分,因为摩擦学问题广泛存在于动力系统、传动系统和机械加工过程等中。摩擦学对加强创意设计和提升产品性能有很大的帮助,并且在工程中可以提供有效的解决主要技术问题的措施与对策。

在机械设计中,摩擦学设计的作用主要体现在两方面:一是在摩擦过程中减少过度摩擦和

磨损;另一个是通过摩擦传递运动和力。摩擦学是研究两个相对运动和相互作用物体摩擦、磨损和润滑的科学和技术。这个词是1966年在希腊语"tribos"的基础上创造出来的,"tribos"的意思是摩擦。摩擦学不仅与我们的日常生活如洗脸、刷牙,密切相关,而且还与众多领域的技术和科学,如航天(空间摩擦学)和地震(地震摩擦学)相关。它的范围很广泛,是一门涵盖数学、化学、物理学、材料学、机械工程等的交叉学科。摩擦学设计的任务是根据力学、材料学和表面工程学研究创新原理、新产品功能设计和实现低摩擦、高耐磨性,以达到使产品节能、寿命长和工作性能更好的目的。

根据摩擦学设计理论,机械元件之间的接触条件大致可分为干接触、边界接触、混合接触、薄膜润滑、弹性流体动力润滑(EHL)、液体动力润滑和流体静力润滑,如图14.1所示。图中,混合润滑是几个接触条件共存的状态。表13.1给出了典型的薄膜厚度、形成机理和上述接触条件的应用。

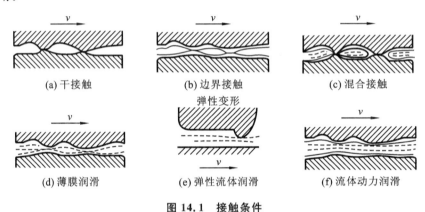

图 14.1 接触条件

表 14.1 接触状态的特征

接触状态	薄膜厚度	油层形成机理	应　　用
干接触	1～10 nm	氧化层,气体吸附层	应用于没有润滑或自润滑的摩擦副
边界接触	1～50 nm	润滑膜的形成是由于润滑剂和金属表面发生了物理或化学反应	应用于低速重载情况下的高精度摩擦副
薄膜润滑	10～100 nm	通过两个相对运动物体的相互作用产生的动力效应而形成	应用于低速情况下的高精度线/点接触摩擦副,不包括高精度的滚动轴承
弹性流体动力润滑	0.1～1 μm	同上	应用于中高速情况下的高精度线/点接触摩擦副,不包括齿轮和滚动轴承
流体动力润滑	1～100 μm	同上	应用于中高速情况下的面接触摩擦副,不包括滑动轴承
流体静力润滑	1～100 μm	由于外部压力把润滑剂传送到两个接触表面之间的缝隙而形成	应用于低速情况下的面接触摩擦副,不包括滑动轴承或导轨

3. 计算机辅助设计

计算机辅助设计(CAD)也称为计算机辅助设计和制图(CADD),是使用计算机系统来协助创建、修改、分析或优化设计。计算机辅助制图介绍使用计算机软件建立一份技术图样的过程。CAD 软件用于提高设计者的生产率,提高设计质量,通过文件促进交流,并创建一个制造数据库。CAD 的输出往往是用来打印或加工操作的电子文件。CAD 软件可使用矢量图形以描述传统图样的,也可产生光栅图形以显示设计产品的整体外观。

CAD 经常涉及的不仅仅是形状。根据应用习惯,与手工绘制的技术及工程图样一样,输出的 CAD 图必须根据具体应用传达信息,如材料、工艺、尺寸、公差。CAD 可以用来设计二维(2D)的曲线图形或三维(3D)的线、面和体。

CAD 是一种重要的工业技术,广泛应用在汽车、造船、航空航天等行业,以及工业和建筑设计、整形和其他许多领域。CAD 也被广泛用于制作电影里面的电脑动画特效、广告和技术手册。计算机现在的普及性和能力,意味着甚至香水瓶子和洗发水喷头设计,所采用的技术都是 19 世纪 60 年代的工程师闻所未闻的。由于具有巨大的经济意义,CAD 成为推动计算几何、计算机图形学(硬件和软件)和离散微分几何研究的主要技术。

在 CAD 中,许多命令可用于绘制基本的几何形状,例如圆、多边形、圆弧、椭圆等。

4. 可靠性设计

可靠性工程是进行可靠性的研究、评估和生命周期管理的工程技术。可靠性是一个系统及其组成部分在规定条件下和规定时间内完成要求的功能的能力。可靠性工程是系统工程的一门分支学科。可靠性用失效概率、失效频率来衡量,或从可行性的角度出发,用可靠和可维护的概率来衡量。可维修性和维修往往是可靠性工程的重要部分。

可靠性工程和安全技术是密切相关的,因此它们使用同样的分析方法,并且可能需要相互引用。可靠性工程的重点是系统停机造成的成本、浪费备用件和维修设备的成本、人员和保修成本。安全工程的重点通常不是成本,而是保护生命和自然,因此只涉及特别危险的系统失效模式。

5. 可持续设计

可持续设计(也称为环境设计、绿色设计、环保设计、生态设计等)是关于设计物理对象、环境建设和服务的理念,以符合社会、经济和生态环境的可持续发展的原则。

可持续设计的意图是"通过成熟的、敏感的设计完全地消除对环境的负面影响"。可持续设计表现为不需要非再生资源,对环境的影响最小和建立了人与自然环境的联系。

除了"消除负面环境影响",可持续设计必须创建有意义的、能转变人们日常行为的创新项目。经济和社会之间的动态平衡,意味着建立用户和对象/服务之间的长期合作关系,并最终要尊重和考虑环境和社会的差异。

6. 设计方法学

设计方法的目标是获取关键的见解或独特的原理,从而找到更全面的解决方案,以使用户对他们依赖的产品、服务、环境和系统有更好的体验。"见解"在这种情况下意味着通过设计方法明确和深入的调查研究,从而直观地把握事物的内在本质。

设计方法学是研究产品设计规律、设计程序及设计思维和工作方法的一门综合性学科。设计方法学以系统工程的观点分析设计的战略、进程和设计方法、手段的战术问题,在总结设计规律、启发创造性的基础上促进研究现代设计理论、科学方法、先进手段和工具在设计中的综合运用,对开发新产品、改造旧产品和提高产品的市场竞争能力有积极的作用。

第15课　计算机集成制造

面对全球化的竞争,制造工业力求不断减少产品的成本以保持竞争力。此外,还需要持续提高产品的质量和性能。另一个重要的要求是准时交货。在全球范围的外部采购和很长的、穿过几个国界的供给链背景下,不断地缩短交货时间确实是一项艰巨的任务。

为了满足上述要求,制造业在机械化的生产线上越来越多地采用了自动化设备和控制技术。自动化的发展促使各工业领域发展自动化"孤岛",如计算机辅助设计、计算机辅助制造、柔性制造系统、自动存取系统、机器人技术、全面质量管理、制造资源计划、办公自动化等等。然而,在这些"孤岛"之间的联系曾一度是由手工操作来实现,这就限制了生产率的提高,而生产率的提高是要在整个制造过程中实现的。计算机集成制造可以视为后续技术,这一技术将这些"孤岛"和其他生产自动化成员联合起来以克服这些限制。

计算机集成制造充分利用了数字计算机的功能来改进生产制造。其中的两个功能是:(1)变量和编程自动化;(2)实时优化。通过采用强大的计算机系统来集成生产制造的所有阶段(见图15.1),从最初的客户订单到最终的发货,生产商希望提高生产率,改善产品质量,尽快满足客户的需要以及提供更多的灵活性。这就要求在工厂不同的部门或区段之间共享信息,访问不相容的、不同类的数据和设备,例如,在设计阶段创建出来的产品数据。必须将这些数据从建模软件毫无缺失地传送给制造软件。计算机集成制造采用了无论在何处都可行的通用数据库和通信技术来集成设计、制造和相关的商务功能,不同功能所需要的数据以无缝的方式从一个应用软件被传送到另一个软件。因此,计算机的功能不仅仅是针对不同的、小范围的生产制造活动,而且是针对整个制造系统来开发的。计算机集成制造减少了生产制造中人的参与,因此在制造过程速度快、成本低更不容易出错。

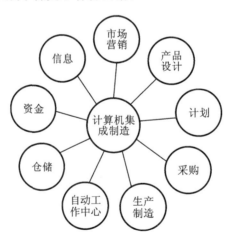

图15.1　计算机集成制造系统的要素

用于产品规划、库存量控制或编制计划的复杂程度较低的计算机系统通常被认为是计算机集成制造系统的一部分。在考察计算机集成制造的实施情况时,会考虑这些因素:产量、企业或个人的集成经验、产品本身的集成水平和生产过程的集成水平。当今,不到1%的美国制造公司实现了全方位地采用计算机集成制造,但是有超过40%的公司正在使用一种或几种计算机集成制造相关技术。一些研究就企业在构成计算机集成制造系统的几项技术上的投资情况对其经营者进行了调查,并通过7分制来衡量(1分＝没有投资,7分＝大力投资)。计算机辅助设计技术得到的平均分最高(5.2),之后依次是数控机床技术(4.8)、计算机辅助制造技术(4.0)、柔性制造系统技术(2.5),自动化物料处理技术(2.3)和机器人技术(2.1)。

要开发一个平稳运转的计算机集成制造系统,有三个主要的挑战。

- 集成来自不同供应商的组元:有时不同的机器(比如计算机数控机床、传送带和机器人)采用不同的通信协议。以自动导引车系统为例,甚至充电时间的长度都可能带来问题。
- 数据的完整性:自动化程度越高,用来控制机器的数据的完整性就越关键。大量来自不同传感器的操作数据和信息,必须利用所获得的有用的知识和资料来及时地收集和处理。计算机集成系统可节约操作机器的劳动力,但它要求有额外的人力来确保对用来控制机器的数据信号有恰当的安全防护。
- 过程控制:计算机可以用来帮助人类操作者操作生产设备,但是必须有能胜任的工程师在一旁来处理一些情况,这些情况是控制软件设计者无法预见的。对操作者来说,要理解显示板、视频和计算机屏幕上的如此多的信号越来越困难了。

第16课　工业机器人

机器人有很多用途。它们被用于勘测火山、开发水下矿床、救火和拆弹、检查核能场所、清洁居室、实验室研究、医学手术、办公邮件递送以及无数其他的任务。今天,大多数机器人在工业生产中得到使用。机器人可以在形式上与人相似,但是工业机器人完全不像人类。最为广泛接受的工业机器人的定义是由机器人工业协会提出的:工业机器人是可重复编程、多功能的机械手,其设计目的是通过可改变的编程动作去移动材料、零件、工具或专用设备,来执行不同的任务。就常规的机器人技术而言,大部分类型的工业机器人都属于机械手(见图16.1)的范畴。

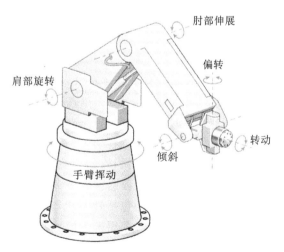

图 16.1　工业机器人及其标准动作

机械手可以分为两部分:(1)主体手臂,这一部分通常由三个与大连杆相连接的关节组成;(2)手腕,由两到三个紧凑的关节组成。机械手的这两个部分有着不同的功能:主体手臂用来使机器人在工作空间内移动和确定零件或者工具的位置,而手腕则用来在工作位置上确定零件或工具的方向。安装在手腕上的是末端执行器,用来执行任务,因而被称为机器人工具。末端执行器可以是焊接头、喷枪、刀具或者带有可开、闭钳爪的钳子,这取决于机器人的具体应用。

1. 工业机器人的特性

- 轴数——到达平面内的任意一点需要两轴,到达空间内的任意一点需要三轴。要完全控制手臂端部(即手腕)的方向还需要另外的三轴(偏转、倾斜、转动)。
- 工作范围——机器人可以到达的空间区域。
- 运动学——机器人中刚性构件和接头的实际排列,决定了机器人可能的运动。
- 承载能力——机器人可以提起的重量。
- 速度——机器人手臂端部能够多快地定位。
- 精确度——机器人可以到达距离指令位置多近的地方。测量机器人的绝对位置并与指令位置相比较,其误差就是精确度的衡量标准。精确度可以通过采用外部感测系统如视觉系统或红外线设备来得到改进。
- 重复性——机器人完全地返回到编程设定点的程度。这与精确度是不同的。当被指示要到达一个由 x,y,z 坐标确定的点时,它(机器人)到达了距离该点 1 mm 以内的位置,这是精确度,精确度可以通过校正来改进。但是如果这个位置被示教并示教信息被储存于控制器内存中,每次机器人被送往那里,它都会回到距离示教位置 0.1 mm 以内的地方,则重复性就为 0.1 mm。

精确度和重复性是不同的量度标准。对一个机器人而言,重复性通常是最重要的指标。

- 动力源——部分机器人使用电动机,其他的使用液压致动器。前者更快速;后者更强有力,而且在诸如喷涂——在这种场合,一个火花就可能引起爆炸——等应用中更有优势。
- 驱动——部分机器人通过齿轮将电动机与关节连接;其他机器人直接将电动机与关节相连(直接驱动)。

2. 机器人编程

控制着机械手的计算机系统必须通过编程来教机器人详细的动作顺序和其他为了完成任务而必须执行的动作。可以通过几种方式来给工业机器人编程。

一种被称为示教编程。这种方式要求用不同的动作来驱动机械手,同时把这些完成指定任务所需要的动作记录在机器人的计算机内存中。示教可以是操作者直接按动作顺序移动机械手操演一遍,也可以是使用控制盒(如示教器,一个手持式控制和编程装置,能够手工将机器人送至所需位置)来驱动机械手按顺序动作。

另一种编程方法要使用一种编程语言,这种编程语言与计算机编程语言很类似。专门的机器人软件可以在机器人控制器或者是与控制器相连的计算机中运行,也可以同时在这两者中运行,这取决于系统的设计。除了有计算机编程语言的很多功能(即数据处理、计算、与其他计算机设备通信、作出决定),机器人编程语言还包括针对机器人控制而特别设计的语句。这些功能包括运动控制和输入/输出。运动控制命令用来为机器人导向,将其执行器移动到某个指定的空间位置。例如,语句"move P1"可以用于引导机器人到达空间中名为 P1 的点。输入/输出命令用于控制信号(来自于传感器和工作间里的其他设备)的接收并且向工作间里的其他设备发出控制信号。例如,语句"signal 3,on"可以用于开启工作间里的一台电动机,这台电动机与机器人控制器里的 3 号输出线相连。

示教器或计算机在编程后通常被断开,机器人接下来就按照安装在它的控制器里的程序来运行。

第17课 污水处理设备

废水通常是通过污水管道系统收集并输送到污水处理厂的废弃液体。废水一般分为两大类：生活废水和工业废水。废水处理是从污水中去除悬浮固体、不良化学物质、生物污染物以及气体的过程。这个过程包括初级处理、二级处理和三级处理。初级净化系统的作用是去除固体及悬浮物，为后期的生物处理作准备。二级处理系统利用含有好氧微生物的生物反应器去除溶解在水中的有机物。污水的三级处理或称高级处理系统普遍用于去除污水中某些特殊残留污染物和微量元素，以及在污水被排入地表水系或入海口之前对其进行消毒。另外，正在研究和使用的先进废水处理方法有蒸发、蒸馏、电渗析、超滤法、反渗透、冷冻干燥、冻结解冻、气浮以及应用土壤处理的方法，特别值得强调的是使用天然或人工湿地处理污水。下面介绍几种典型的污水处理设备。

沉淀池　颗粒受到重力或离心力作用时，会沿作用力的方向定向运动。如果是重力沉降，颗粒将因重力而沉降到池底，并形成污泥。在初级处理阶段，当污水流过沉淀池时，在污泥沉降的同时，污水表面漂浮的动植物油脂也被刮除（见图17.1）。初级沉淀池装有机械刮泥板，可以连续地将污泥推入池底的料斗，料斗中的污泥被泵送到污泥处理设备。

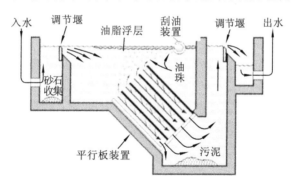

图17.1　沉淀池工作原理图

生物反应器　有许多不同种类的生物反应器应用于环境工程实践。生物反应器包括悬浮生长和固定膜生长系统两大类。悬浮生长净化系统中含有活性污泥，由活性污泥菌团与污水混合，对有机物进行降解；固定（或附着）膜净化系统包括滴流式生物滤器、生物塔和生物转盘（筒）等，生物膜附着在固体介质上，对流过其表面的污水进行净化。选择生物反应器时主要考虑以下影响因素：废水的物理和化学特性、污染物的浓度、污水含氧量、处理效率和系统可靠性要求、反应器工作的气候条件、整个处理系统中所包含的不同生物处理单元的数量、系统操作的技术和经验以及不同结构反应器的施工和运行成本等。图17.2介绍了几种常用的生物反应器。

固定填料床反应器　固定填料床是最常见的固定膜反应器。以前主要采用大石块作填料，但目前通常采用塑料颗粒或用豌豆大小的石块作填料，与采用大石块相比，采用这种填料时反应器单位体积的重量较轻，同时具有更大的比表面积和孔隙率。滴流式生物滤器和生物塔是人们一般都知道的固定填料床反应器。在反应器中污水均匀地分布在床体上，并缓慢地流过介质表面，使固定填料床生物反应器有推流式特性。填料之间的空隙可允许空气进入，为反应器中的微生物提供更多的氧气。

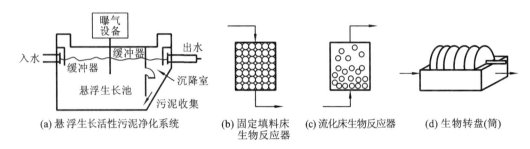

图17.2 典型生物滤池结构

流化床反应器 在足够高的上升流速的承托下,附着微生物的固体介质在反应器中呈悬浮状态。流化床常用的载体有沙粒、活性炭颗粒、硅藻土或者其他抗磨损的小颗粒。污水的入流速度能否保证载体处于悬浮状态,取决于附着在载体上的生物量的多少。

生物转盘反应器(RBC) 生物转盘是另一种固定膜反应器。与流化床反应器相似,这种反应器具有良好的混合特性和传质特性。转盘和装有塑料颗粒的转筒被安装在中心转轴上。当转轴旋转时,暴露在空气中的盘体上的生物膜直接与空气接触并吸收氧气,浸没在污水中的盘体的生物膜对污染物进行分解净化。废水可以从生物转盘反应器的一端进入然后垂直流过反应器,因此产生推流式特性。或者,污水沿着反应器的长度方向均匀地流入,使整个系统完全混合。

曝气设备 曝气设备用于污水曝气,需要生物净化过程,还需对污水进行混合,使固体颗粒处于悬浮状态而得到更有效的净化。曝气设备类型有很多,目前污水处理中主要采用两种,即机械搅拌式表面曝气设备和水下空气或纯氧扩散设备(见图17.3)。对传统的活性污泥污水处理厂而言,曝气系统的能量消耗占总能耗的45%~60%。因此曝气系统运行状况直接影响污水处理厂的投资费用和运行成本。

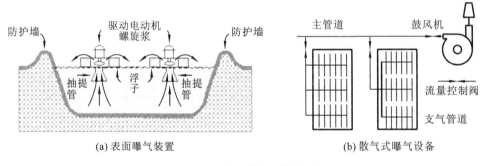

图17.3 常用曝气设备类型

膜过滤设备 膜过滤是流体通过微孔膜去除污染物的过程。常用的膜过滤方式包括微滤(MF)、超滤(UF)、纳米过滤(NF)和反渗透(RO)。微滤是一种使用特殊膜过滤的方法,这种膜的孔径在 $0.03\sim10~\mu m$ 之间,微滤因滤膜孔径大小得名。微滤用于去除粒径在 $1~\mu m$ 以上的固体颗粒。超滤是一种用来去除水中非常微小颗粒的过滤方法。超滤膜的孔径在 $0.01\sim0.001~\mu m$ 之间,因此能够去除大部分的细菌、病毒、大分子物质以及聚合物分子等。超滤通常用于净化饮用水等对水质标准比较高的水体。纳滤膜的孔径约为 $0.001~\mu m$,污水通过这种膜时需要的压力比微滤和超滤更高,其压力接近 600 kPa(90 psi),有时高达 1000 kPa(150 psi)。纳滤可以去除水中所有的动植物胞囊、细菌、病毒以及腐殖质等,同时还可以用于去除饮用水中的化学污染物和杂质等,也可以用于盐水的淡化。反渗透可以有效去除水中的无机污染物,也可以去除放射性镭、天然有机物、杀虫剂、动植物胞囊、细菌和病毒等。在反渗透系统中,对浓缩液施加压力使水分子透过膜到达新鲜水的一边。一般情况下,膜是由合成的聚合物制成的,其他如陶瓷膜和金属膜也可使用。目前几乎所有用于饮用水净化的膜都由聚合物材料制成,这是由于聚合物比其他材料更经济。常见的膜结构有中空纤维膜、卷式膜和筒式膜组件(见图17.4)。

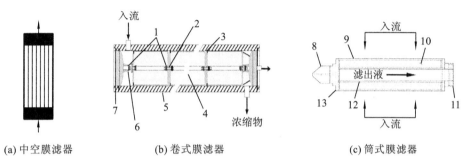

图17.4 常用的膜过滤器结构形式

1—O型圈;2—内连接件;3—盐水密封;4—卷式膜;5—压力容器;6,8—端部适配器;
7—端部密封;9—外壳;10—过滤介质;11—底部适配器;12—滤芯;13—端帽

大多数用于饮用水净化的中空纤维膜组件是由微滤膜或超滤膜制成的,用于去除水中的特殊物质。顾名思义,中空纤维膜组件由中空纤维膜构成,纤维膜是长并且很细的管。纤维膜可以按不同的排列方式捆扎而成。常用的中空纤维膜组件的制造方法是:首先将纤维膜沿其长度方向捆扎在一起,再将两端分别固定在树脂底座中,最后将捆扎和固定好的纤维膜安装在一个压力容器内,这样就组装完成了一套中空纤维膜组件。组件一般为垂直结构的,尽管水平结构的也可以使用。

卷式膜组件采用半透膜,可有效去除水中的溶解性物质,因此经常与纳米过滤或反渗透工艺联合使用。膜组件的基本单元为三明治结构的,它由多个叫做叶膜的平板膜片卷绕到中心管上而形成。每片叶膜由两张相背的膜片构成,两张膜片由纤维质的渗透支承隔开。叶膜的三条边由胶密封,第四条边与中间通道集水管相连并且密封。一个卷式膜组件的直径有8 in (1 in=25.4 mm),它可以容纳20片叶膜,每片叶膜之间用塑料网片隔开。塑料网片是污水的进水通道。

图17.4(c)是典型的筒式膜滤器示意图。筒式膜滤器的组装过程为:首先将平整的膜材料放置在分别与原液和滤出液接触的两个支撑层之间,再将三者褶皱叠起以增加滤膜的表面积,将褶皱的膜过滤材料安装在滤筒与外壳之间,用黏结剂或通过加热将膜组件的两端密封好。端部连接套管采用双O型圈机械密封,保证筒式膜组件与外壳的密封良好。

第18课 再制造工程

由于产品和制造过程每年产生超过60%的废物,日益严格的法律要求其降低对环境的影响。例如,生产者有责任回收使用过的产品,减少填埋。这些压力,再加上全球工业活动的激烈竞争,使得企业改变对产品设计的态度。公司必须设计寿命长的、在报废时容易回收的产品,必须考虑废旧产品加工的商业潜力,以挖掘零部件的剩余价值。

再制造是将使用过的产品在其功能上"整旧如新"的过程,此过程不仅可获利,而且比常规过程对环境危害作用小,因为它能减少垃圾填埋和生产过程中原材料、能源及具有专门技术的劳动力的消耗量。再制造的主要障碍包括消费者的接受程度低、缺少再制造刀具和技术,以及许多在用的产品可再制造性差。这些障碍来源于缺乏再制造知识,包括对其定义的理解不全面。这些词汇,像修复、重置和再制造经常作为同义词。因此,顾客对再制造产品的质量缺乏信心,对购买再制造产品持谨慎态度。同时,设计者在工作中考虑产品寿命末期问题,如产品的再制造时可能欠缺相关知识,因为传统的设计主要关注功能和解决环境问题的成本。另外,要解决企业转型中的相关问题,需要对技术、企业创新与社会发展水平的关系,以及这种创新的过程进行研究。对制造管理,特别是初期转型的变化的决策过程,需要进行评估。再制造的方式体现了大多数制造商在决策过程中的基本转变。从历史上看,废弃物管理方法的选择基于对可衡量的成本和经济效益的纯经济的分析。这种做法忽略了大量定性因素的影响,在决策过程中,这些因素影响着合适的技术的选择。另一方面,再制造是一种复杂的、多学科、多功能的实践方式,可能使工业中大量的废物最小化技术变得可用,这些技术包括改变产品、改变生产过程中使用的材料、改变操作和回收方法等。它需要各个设计和以数据为基础的活动,例如环境影响分析、数据和数据库管理、优化设计等相互协调。现在有一个共识,就是决策受不可预测性、风险和不确定性的影响。例如,某些技术的影响能够准确地计算吗?怎样区别和计算经济、社会和环境等不同方面的风险?某些实践在一个企业取得成功,在另一个企业为什么会失败?

再制造工程是以产品全寿命周期理论(见图 18.1)为指导,以废旧产品性能跨越式提升为目标,以优质、高效、节能、节材、环保为准则,以先进技术和产业化生产为手段,来修复、改造废旧产品的一系列技术措施或工程活动的总称。简言之,再制造工程是废旧产品高技术修复的产业化。实践证明,再制造可使废旧产品中蕴涵的价值得到最大限度的开发和利用,是废旧机电产品资源化的最佳形式和首选途径,是节约资源的重要手段。对废旧机电产品进行再制造是发展循环经济、建设节约型社会的重要举措。

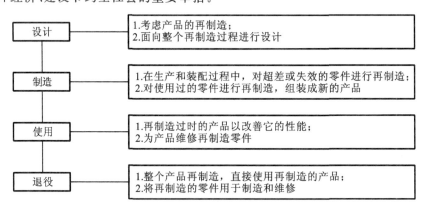

图 18.1 再制造的产品寿命周期

在机电设备的使用过程中,零部件会产生磨损、变形、断裂、蚀损等形式的失效,设备的精度、性能和生产率将下降,从而导致设备发生故障、事故甚至报废,因而需要及时地对其进行维护和修理。在维修过程中,所有方法的目的都是用最短的时间、最低的成本来有效地消除失效,使设备得到更有效的利用。修复工艺能使零件再生,达到这个目的。有一些常见的修复方法,例如机械修复、焊接、热喷涂、电喷涂、胶结、刮研等等。下面以氧乙炔焰喷涂工艺为例来阐述修复工艺。

氧乙炔焰喷涂工艺

氧乙炔焰喷涂的简图见图 18.2。氧乙炔焰喷涂的过程包括表面处理、预热、喷涂基体粉末、喷涂工作层以及加工喷涂层。

除了铜和钨基材料，所有的普通钢、不锈钢、硬质合金钢、氮钢、镍、铬合金和铸铁都能用来喷涂。

1) 喷涂前的准备

喷涂前的准备过程包括工件清洁、表面处理、表面强化和预热。

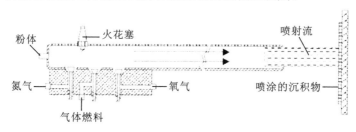

图 18.2 火焰喷涂简图

工件清洁的主要对象是被喷涂的部分，以及它们附近的油、腐蚀层和氧化层。用火焰焙烤去除表面杂物，以保证结合的质量。

表面处理的目的是去掉疲劳层、碳化层、工件表面的涂层和表面损伤，调整不平的磨损表面，保留涂层厚度，这些工作决定了预处理的质量。表面处理常用的方法有车削和磨削加工。

表面强化是对喷涂表面的强化处理，以提高喷涂层和基体的结合强度。常用的表面强化方法有喷砂和电火花强化。当然，也可以用车削、磨削和滚花的机加工方法。车削加工强化处理用来加工螺距为 0.3~0.7 mm，深度为 0.3~0.5 mm 的螺纹。

预热的目的是去掉表面吸收的水分，减少冷却过程中的收缩应力，提高结合强度。

2) 喷涂结合层

预热以后应立即在工件上喷涂结合层来提高工作层和基体的强度。喷涂结合层对于薄工件和喷砂时易变形的工件特别适用。

结合层的厚度为 0.1~0.15 mm，喷涂的距离为 180~200 mm。

3) 喷涂工作层

喷涂结合层以后应立即在工件上喷涂工作层。工作层的质量取决于送粉速度和喷涂距离。送粉速度快，送粉量增加，涂层质量下降；送粉速度太慢，效率下降。喷涂距离太近，加热粉体的时间不够，工件的温度较高；距离太远，合金粉体的速度和温度下降。工件表面移动的线速度为 20~30 m/min。在喷涂过程中，粉体的喷射方向与喷涂表面垂直。

4) 喷涂后热处理

喷涂后工件缓慢冷却。由于喷涂层多孔，要喷涂防腐液以防止涂层被腐蚀。通常情况下，在涂层的表面喷涂油漆和环氧树脂。

当涂层的尺寸精度和表面粗糙度不能满足要求时，可进行车削或磨削加工以达到要求。

第 19 课 矿山冶金设备

矿物加工是在采矿后大多数矿石必须经过的第一个工序，目的是为提取冶金工序提供富集度更高的原料。其主要的操作是粉碎和精选,但在现代矿物加工工厂里,还有其他重要的操作,包括取样、分析和脱水。

在采矿后,大块的矿料被压碎和/或磨碎以获得足够小的颗粒,使每个颗粒要么基本上是有用的,要么基本上是废料。下一步的处理就是精选。在现代冶金中,精选工序是非常重要的,这是由于人们希望提高冶金生产的效率,也由于金属冶炼业的发展要求利用更低等级的矿石。不经过精选而直接对这些矿石进行冶金处理通常是不经济的,有时候甚至是不可能的。浮选、重选、磁选和电选是最普遍采用的精选方式。在精选中,矿石被放置在机器中压碎,通过摇动,含有金属的重颗粒由于重力差别而与较轻的岩石颗粒分离开来。浮选工序用来处理90%以上的非铁金属和稀有金属矿石。在某些情况下(例如当金、银或者偶尔有铜以自由态出现,即在砂或岩石中未化合时),单独采用机械的或者选矿方法就可以有效地获得相对纯净的金属。报废的矿石被冲洗掉,或通过筛选和重力作用分离开;富集矿石接下来要经过不同的化学工序来处理。

在经过矿物加工的精选工序之后,要对金属矿物进行提取冶金,在这一过程中金属元素从化合物中被提取出来,经精炼去除杂质。由于自然界中几乎所有的金属都与其他元素化合在一起,因此需要用化学反应来将它们分离出来。这些化学工艺分为火法冶金、电冶金和湿法冶金。

火法冶金,或使用热量来处理矿石,包括焙烧和熔炼。金属化合物常常是相当复杂的混合物,而且它们通常不属于那种使用简单经济的工序就能从中提取出金属的化合物类型。因此,进行提取冶金时,在把金属元素与化合物的其他成分分离开来之前,通常必须先把化合物转变成可以更容易被处理的类型。一般用来完成这项工作的预先处理方式就是焙烧,在焙烧过程中化合物在恰好低于它们熔点的温度下被转化。焙烧有几种不同的类型,每一种都是为了引发一个特定的反应,并且产生一种适合于后续特定加工操作的焙烧产品(焙烧产物)。硫化物矿石通常在空气中加热。矿石中的金属与空气中的氧气结合形成氧化物,矿石中的硫也与氧结合形成二氧化硫气体而消失。接下来用还原剂处理金属氧化物。焙烧过程可以在专门的焙烧炉中进行。沸腾炉(见图 19.1)由于容量大和效率高而被广泛采用。氧化、硫酸化、挥发焙烧都可以使用沸腾炉。焙烧炉的内衬采用耐火材料,有竖直的圆柱钢壳。炉底带栅格,可使足量的空气通过而进入炉内,使细固体给料颗粒保持悬浮状态,从而使气体与固体接触良好。矿

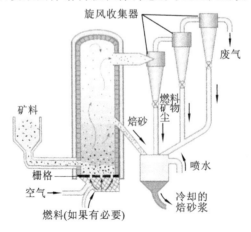

图 19.1　沸腾炉

物料可以是干的,也可以是水的悬浮液,通过下部管道进入焙烧炉的湍流层区域。焙烧产物可以由侧面的溢出管道排出。

熔炼是通过把岩石加热到高于金属熔点的温度,将金属元素从它的化合物中"解放"出来,形成不纯的熔融金属,再将熔融金属与炉料中的废岩石部分分开的过程。如果矿石是氧化物,就与还原剂一起加热,如用碳(以焦炭、煤的形式存在)做还原剂时,矿石中的氧与碳结合,形成二氧化碳气体而被除去。矿石中的废料称为脉石,通过一种称为造渣剂的物质来去除,受热时造渣剂与脉石结合形成熔融的物质,称为炉渣。炉渣比金属轻,浮在金属的上面,可以被撇去或排出。造渣剂的使用是根据矿石的化学性质来确定的。石灰石通常与硅质脉石一起使用。

电冶金包括采用电解作用制备某些活性金属,如铝、钙、钡、镁、钾和钠。电解是指将熔融的金属化合物(一般是氯化物)接通电流,使金属在阴极聚集。

湿法冶金由浸出操作和电解冶金组成。在浸出操作中,利用水溶剂将金属化合物有选择性地从矿石中溶解出来;在电解操作中,在溶液中通以电流,使金属离子沉积在电极上。例如,用稀硫酸处理某种铜的氧化物和碳酸盐矿石,形成溶于水的硫酸铜,金属通过电解水溶液的方法被回收。

提取之后通常是精炼,在精炼工序中,通过火法冶金、电解或化学手段,杂质的含量得以降低或被控制。火法精炼通常是使杂质在高温熔池中氧化而去除。电解是使电解池中的一个电极溶解,更纯的金属在另一个电极上沉积。化学精炼包括把金属转化成蒸汽而凝结或者使金属从水溶液里沉淀析出。

第 20 课 可持续产品设计

1. 引言

可持续设计(也称绿色设计或生态设计)是指在设计产品时专门考虑其在整个生命周期中对环境影响的设计方法(见图 20.1)。可持续设计具有特定的设计框架,包括考虑环境问题、相关分析及综合方法的应用,是对传统设计和制造方法的挑战。

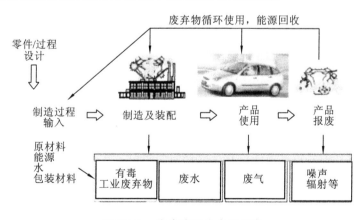

图 20.1　汽车产品生命周组成

过去,在产品设计和制造阶段往往忽视对环境的影响,危害废弃物一般被随意丢弃,不考虑可能对环境造成的危害等。另外,能源利用效率低,导致运行成本高昂。在产品的材料生产、制造和销售过程中,浪费现象随处可见。消费者丢弃的废品一般只有少量被再制造或循环使用。认识到这些问题之后,环境工程技术的研究开始转向对已造成污染的治理(环境修复)和当前污染治理(废物处理)。末端治理技术至今在很多场合仍然需要,但是如果改变产品的设计方法,对减少环境污染和降低成本更为有效。具体的操作方法有以下几种。

溶剂取代 用良性的溶剂取代过去单一使用的有毒溶剂,比如使用可生物降解的溶剂或者无毒溶剂。水基溶剂优于有机物溶剂。

技术革新 诸如研发更节能的半导体材料或汽车引擎之类。例如,"能源之星"计划就是规定了计算机、打印机和其他电子设备的最大能耗标准,能耗符合该标准的产品可以被贴上"能源之星"标签。同样,"绿色照明"计划的目的也是开发耗电更少的照明产品。

有毒废弃物回收 避免有毒污染物排入环境和产生新的污染物。例如,镍镉充电电池中的镉和镍可以被全部回收,并应用于其他地方。

可持续产品设计面临的挑战是如何改变传统的产品设计方法和制造工艺,并将环境因素系统、有效地结合到可持续产品设计中。这要求改变现有的设计方法。产品设计方法的革新是一项艰巨任务,因为设计人员需要面对设计目标冲突、不确定性、设计时间和成本效益等诸多问题。必须将环境因素实质性地融入产品设计的复杂过程中。为了满足这一要求,必须确立可持续产品设计的基本理论和创新设计方法。除了对传统的产品功能、造价、性能和上市时间等方面的考虑,工程师还必须顾及产品的可持续性。工程师必须从减少能源消耗、采用无废生产工艺、减少原材料使用量以及回收再利用报废产品等方面考虑产品的设计,也就是以产品的生命周期为基础开展产品设计(见图 20.2)。当然,所有这些都必须在利益相关者的参与和采用创新技术和方法的条件下实现。

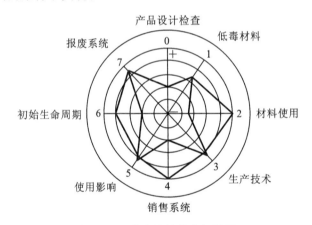

图 20.2 生态设计轮式规划图

2. 可持续设计的目标

关于可持续设计的目标目前还没有达成普遍的共识。某些人认为可持续设计的唯一目的就是降低成本。根据这种观点,生产过程或产品产生的任何废弃物都是机会。有些人将可持续设计定义为特定策略,例如通过回收节省原材料,并且采取的所有措施都是为了达到这一目的。还有一种观点认为可持续设计是为了解决某项特定的环境问题或环境因素,比如全球气候变暖问题、空气污染问题,不考虑其他环境影响。可见,上述每一种观点都有缺陷。

考虑到资源和生态健康,可持续设计的社会学目标是确保未来社会的可持续发展。根据可持续发展要求,可以提出三个可持续设计总体目标:

- 减少或者以最小量使用不可再生资源;

- 管理可再生资源以确保可持续性；
- 减少有毒和其他有害物排放到环境中的量，包括减少温室气体排放，这是终极目标。

可持续产品设计的目标是采用最经济有效的手段实现上述目标。

3. 可持续设计策略

可持续设计的核心是考虑设计决策的系统效应。设计一个新产品时，在材料供应、制造、使用和处理阶段都要考虑可能产生的环境污染问题。可持续设计的策略包括以下内容。

生命周期的评价（LCA） 生命周期的评价是评价产品"从摇篮到坟墓"即从概念设计到报废的整个生命周期（如图 20.3 所示，包括原材料加工、制造、销售、使用、维护和保养、处理或再循环）对环境影响的技术。通过以下几种手段，生命周期评价有助于避免片面地看待环境影响问题：

- 编辑有关能源、材料输入和环境排放的详细目录；
- 根据确定的输入和排放评价潜在的影响；
- 解释能帮你作出更明智的决策的结果。

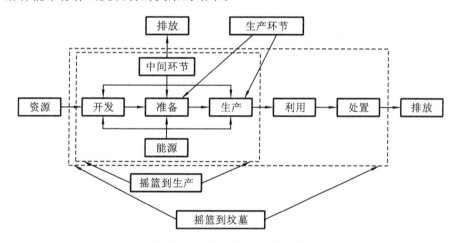

图 20.3 产品的生产边界条件

物流循环跟踪 物流循环跟踪技术用来确定产品生产过程中材料的用量及其使用的时间和地点情况（见图 20.4）。例如，由报废的汽车等机械产品中回收废旧钢铁，经过冶炼后再重新利用是一个闭式物流循环。然而有些材料尽管将来可能被回收或者再利用，在报废后还是被填埋了。材料的闭式循环（重新用于同一用途）和开式循环（用于不同用途，特别是用于较低质量要求的场合）是有区别的。弄清楚物料的循环边界和一些不确定因素，对物流循环跟踪十

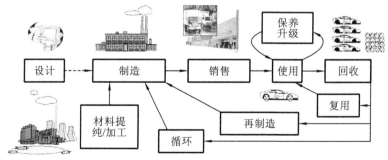

图 20.4 产品生命周期中材料循环利用流程图

分重要。

3R 资源管理技术　3R 技术是指资源的减量化(reduce)、再利用(reuse)和循环(recycle)，是环境意识的基础，能促进能源、材料消耗的减少，从而有利于环境。资源减量化是指在产品的设计、制造、采购等过程中，减少原材料(包括包装材料)和有毒材料的用量，减少废物产生量。再利用是指重复使用产品，其中包括传统意义上的对同一功能的重复使用和利用产品的其他功能。循环是指将废物转换为有价值的资源。将收集的各种用过的瓶子、罐头包装和废旧报纸等投放到路边或者收集设备中，仅仅是增加经济收入、保护环境和有益于社会发展的第一步，这些效益积累不仅有利于当地也有利于全球。

4. 可持续设计方法

在设计新产品时，应该将材料供应、制造、使用和处理与环境污染联系起来。设计和制造可持续产品要求具有相关的知识、生产技术和激励政策。可持续设计方法应该比较容易理解并且被快速地使用。可持续设计的特殊方法包括以下几种。

物料平衡分析　物料平衡分析是针对某一工厂、制造车间或生产单元，对输入和输出的物料和能源情况进行分析、核算的过程。从理论上讲，物料平衡的基础是对系统输入量、产品库存和输出量(包括产品、废弃物和耗散量)的测定。实际上，所有这些需要的数据很少用得上，甚至并不一致。

绿色指数　绿色指数或排名系统用于对影响环境的各种因素进行简单的等级划分。产品设计者和决策者可以比较几个备选方案(材料、加工等)的绿色指数值，从中选出对环境影响最小的一个，从而生产对环境污染少的产品。

基于拆卸和循环的设计(DFD/R)　DFD/R 是指所开发的产品在报废后应该容易拆卸以保证后续的循环和零部件复用。然而，通过物理拆卸得到有价值零部件和材料的花费往往超过材料本身的价格。单纯追求减少产品拆卸时间(由此降低拆卸成本)，往往会亏本。DFD/R 设计软件能够帮助分析产品潜在的拆卸途径，确定最佳拆卸工艺，消除拆卸过程中可能遇到的障碍。

风险分析　风险分析是一种确定产品对环境的不同影响发生的概率的方法，例如，通过估计污染物的排放量和排放种类以及对生态和人类可能的损害等，对污染物排放的毒性风险进行评估。

材料选择和标签指示　材料选择索引可以为设计者提供环境友好材料明细。指示标签一般贴在材料或产品上，提供与处理方法和废物管理等相关的原料的含量内容。将标签嵌入到材料上比用黏合剂贴到材料上要好。

开发和营销可持续产品是实现资源高效利用、环境保护以及可持续经济发展的具体措施。可持续产品意味着资源利用更有效，气体排放量和废物产生量减少，同时污染控制和环境保护的社会成本降低。可持续产品在降低成本(减少材料需求、减少处理费用和减少环境污染治理费用)方面具有很大潜力，并且还会通过增加销售和出口数量以增加盈利，因此，可持续产品设计不仅能造福当代，也有利于后代的生存和繁荣发展。

第 21 课　汽车工程

自二十世纪初开始，汽车进入几乎每个人的生活。汽车是一种可以自动行进、由内燃机驱动的交通工具，用来在路面上运输乘客、货物。机动车有不同类型，包括轿车、公共汽车、卡车、有篷货车和摩托车,其中以轿车最受欢迎。"automobile"(汽车)这一术语源自希腊语"autos"(自己)和拉丁语"movére"(移动),是指汽车能够自己行进。一辆汽车有驾驶座位,几乎都至少有一个乘客座位。今天,汽车是全世界交通运输的主要工具。

汽车的主要组成部分包括发动机、底盘、车身和电气系统。

发动机是一个动力装置或者一个电动机,提供能量来驱动汽车。在大多数的汽车发动机中,由燃料的燃烧形成压力,这些压力通过活塞和连杆传给曲轴,并使曲轴运动,产生扭矩。由发动机产生的扭矩通过传动系传输给车轮来推进车辆。要让发动机工作需要许多系统:润滑系统,用来减少摩擦并防止发动机磨损;冷却系统,用来使发动机的温度保持在安全界限内;发动机还需要燃料系统来提供适量的空气和燃料;空气和燃料的混合物必须在恰当的时间由点火系统在汽缸内点燃;最后,还需要电气系统开动电动机,电动机再启动发动机,并且为发动机附件提供电能。

底盘由车架、悬挂系统、传动系统、转向系统和制动系统组成。车架通常由若干方钢或槽钢构件焊接或铆接在一起形成,最终的轮廓要有足够的强度来支撑车身和其他部件的重量。悬挂系统的功能是吸收由车轮在不规则路面上下运动所产生的振动。弹簧、连杆组、减振器组成了车辆的悬挂系统。对振动的吸收和阻尼保护了乘客,使其免受由振动带来的不适。传动系用于将发动机制造的能量传给车轮。它由离合器(在带有手动变速器的车上)、变速器(一个齿轮系统,用于增加发动机的动力以驱动汽车)、驱动轴、差速器和后桥组成。转向系统用于改变车辆的方向。对任何转向机构的主要要求都是转向准确并容易操控,转弯后前轮应该有回到笔直向前位置的倾向。在转向系统中,用转向齿轮机构来增加由驾驶者提供的转向力。转向系统使驾驶者不需要费太大的力就能使车辆轻易地转向。

制动器用来减速或是让正在行进的车辆停下来。制动系统对乘客和路人的安全而言是非常重要的。制动系统可以采用机械或液压方式。目前使用的大部分的制动系统都是属于液压型的。制动器是液压式的,所以失效是缓慢泄漏,而不是突然的电缆断开。所有的制动器都由旋转的和静止的两部分组成。可以通过不同的方式使两部分接触,这样就可降低车辆的速度。所有的车辆必须安装至少两个独立的系统,它们曾分别被称为运行制动系统和应急制动系统,现在通常指的是脚踏式制动器(脚刹)和驻车制动器(手刹)。在大部分轻型车辆上,液压式脚踏式制动器可控制所有轮子,手动操控的机械式制动器仅控制后轮。手动制动系统的一般用途是在停放车辆时控制汽车。这两种制动系统是独立设计的,因此一旦一个系统失效,另一个系统仍旧可运行。

车身的主要作用是为驾驶人员和乘客提供舒适的座位,并保护其免受风吹雨打。所提供的舒适程度取决于汽车的类型和它的价格。最初的汽车车身和安装着座位的平台没有什么差别,后来逐步地发展到车厢封闭,拥有车顶和车窗。现代汽车车身是由在巨大的冲压机下成形的钢板所建造成的。大部分的车身组成部分被焊接在一起,形成一个轻的不会发出噪声的部件。对所有的钢表面都必须进行处理和喷漆,这主要是为了保护表面,避免其生锈腐蚀,其次是为了改进外观。由不锈钢、镀铬黄铜或塑料制造的外部装饰,与简单着色的表面形成对照,修饰了车身,吸引了眼球。汽车也可以在车身风格的基础上进行分类(见图21.1)。

电气系统提供能量来开动一个启动电动机并为所有附件提供能量。电气系统的主要构件是电池、交流发电机、启动电动机、点火线圈和加热器。当点火开关打开时,该系统启动发动机。电气系统用于制造火花以点燃压缩的空气-燃料混合气体,也用于开启前灯、指示灯、制动灯、停车灯、雨刷,如果车内安装了空调、收音机、盒式录音机等也能将它们开启。电池提供电

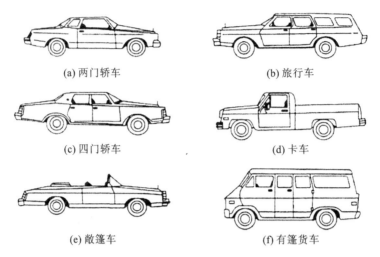

图 21.1 车身风格

能来启动，一旦发动机运转起来，就由交流发电机给全车电气元件供电。交流发电机也给电池充电以补充用于启动发动机所消耗的电能。

第 22 课 工程装备

　　机床是一种用来加工成形金属或者其他硬质材料的机器，通常用来进行车削、钻孔、磨削、切断或其他的成形加工。机床是用一些工具进行切削或成形加工的。所有机床都有一些装夹工件的方法，并让机械零件按既定的方向运动，因此工件和刀具之间的相对运动（即刀具路径）至少在某种程度上是被控制或限制的，而不是完全"即时"或"随意"的。

　　机床这一术语的明确定义对于不同的用户可能也不同。虽然不是所有工厂的机器都是机床，但是可以说所有机床都是帮助人们制造物品的机器。

　　目前机床是典型的靠动力设备驱动（如通过电压、液压驱动或主轴传动）而不是人力，它通过切削或其他某种加工方式来制造机械零件。

　　拉削是一种用齿状的拉刀去除材料的机加工过程。拉削主要有两种类型：直线拉削和旋转拉削。直线拉削工艺更为普遍，它通过拉刀相对工件表面作直线运动来完成切削。直线型拉刀通常用在拉床上，它有时被缩短成钻头形状。在旋转拉削中，拉刀旋转着压进工件来完成轴对称形状零件的加工。旋转型拉刀通常用在一般车床或螺纹车床上。在这两种车床上，切削都是通过拉刀来实现的，它使切削变得很有效率。

　　拉削通常用于精加工，特别是对那些怪异形状的加工。一般的加工表面包括圆形和非圆形的孔、花键、键槽和平直表面；典型的工件包括中小型铸件、锻件、螺纹机械零件和冲压件。虽然拉刀比较贵，但是对于大批量的生产，拉削通常比其他一些加工更受青睐。

　　拉刀的形状与锯子很相似，不过刀齿高度会使其长度增加，而且，拉刀包含三个不同的部分：粗加工部分、半精加工部分和精加工部分。拉削是比较特殊的机械加工，因为它的进给系统固定在机床上；加工表面的轮廓总是和拉刀的表面轮廓相反。齿升量（RPT），也称为一步或每齿走刀量，它决定了切削量和切屑的大小。可以是工件不动，拉刀相对于工件移动；也可以是拉刀不动，工件也可以相对于拉刀移动。因为拉刀有很多固定的功能，所以拉削时不需要复杂运动或技术娴熟的工人。拉刀实际上是一个按顺序排列好的单刃刀具的集合,一个接着一个地切削,有点类似于一个成形刀具的多次切削。

钻床(包括卧式钻床、立式钻床和台式钻床)的固定方式可能是立式固定,或者是用螺栓将其固定在地面或工作平台上,可移动的带有磁力的底座可以夹紧被加工的钢铁工件。钻床由底座、立柱、工作台、主轴和钻床头等组成,通常用异步电动机驱动;钻床头部从中心轮毂起有一组手柄(通常是 3 个),轴转动时,它用来控制主轴的竖直移动从而使其平行于立柱的轴线。工作台可以通过齿条和齿轮进行垂直调整,然而,一些老型号的钻床还要依靠操作者来提升工作台,以使其紧固在合适的位置。工作台也可以在一些旋转情况下偏离主轴到垂直于立柱的位置。钻床的大小主要是由摆幅决定的,摆幅被定义为主轴中心到立柱最近边缘距离的两倍。以一台 16 in(410 mm)的钻床为例,其主轴到立柱最近边缘的距离为 8 in(200 mm)。

插齿机是一种用来加工内、外齿轮轮齿的机器,它的命名跟刀具的运动方式有关;刀具在前进的行程中和工件啮合,在返回时和工件分离,同刨床上的摆动刀架一样。

插齿刀也是齿轮形状的,它和需要被加工的齿轮具有相同的节距。然而,加工内齿轮时,插齿刀的齿数必须小于内齿轮的齿数;加工外齿轮时,插齿刀的齿数由插齿机的型号来决定。那些比较大的齿轮,为了使加工成形更方便,其毛坯会先被加工成大致的形状。

旋转插齿机工作时的主要运动有以下几种。
- 切削运动:刀盘主轴和刀具一起作向下直线运动。
- 回程:主轴和刀具向上运动以撤回到它的起始位置。
- 转位运动:刀盘主轴和工作主轴慢速旋转来进行循环进给,旋转速度通过变速齿轮来控制。被加工齿坯与刀具旋转方向相反,其速度根据 n/N 的大小来设定,其中 n 为插齿刀的齿数。

滚齿是一种在滚齿机上加工齿轮、花键和链轮的加工工艺,滚齿机是一种特殊类型的铣床。通过滚刀,一系列的齿或槽在工件上被加工出来。和其他的齿轮成形工艺相比,滚齿的费用相对是比较便宜的,而且精度也很好,所以,它可用于零件的大批量加工。

切齿工艺广泛用于直齿齿轮和斜齿轮加工,而更多其他齿轮是采用滚齿的方法制造的,因为它效率高、费用低。

珩磨是一种用于精密表面研磨加工的加工工艺,它通是过油石,沿着控制路径对工件进行摩擦,主要用于改善零件表面的几何形状,也可以用于改善零件的表面结构。

珩磨主要应用于内燃机汽缸、空气轴承主轴和齿轮的精加工。珩磨头有多种类型,所有类型的珩磨头都具有一个或多个油石,这些油石在工作时被压紧在加工的工件表面上。

在日常应用中,珩磨棒可以用来磨刀,特别是对于厨房使用的菜刀,与粗刃磨相比,这是一种很好的加工方法,具有更好的磨锐性能。

车床是机床的一种,它使工件跟主轴一起旋转,利用刀具来完成诸如切削、砂磨、滚花、钻孔或变形加工等多项操作,这些操作适用于加工关于旋转轴对称的实物。车床的主要部件如图 22.1 所示。

车床主要用于木材车削、金属加工、金属旋压和玻璃加工。它还可以用于陶器的成形,其中最著名设计就是陶轮。装备完善的金加工车床可以用来加工出回旋体、平面和螺纹或螺旋线。装饰用车床可以用来生产很复杂的三维艺术品。材料通过一个或两个中心点固定,其中至少有一个可以水平移动,以适应不同长度的材料。其他装夹方法包括用卡盘或夹头把工件夹紧,并关于旋转轴定位;或采用花盘,用夹钳或止块装夹。

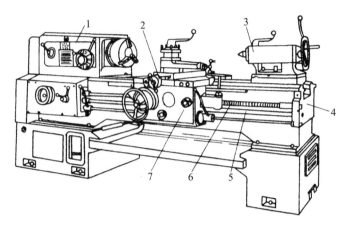

图 22.1 车床的基本结构示意图
1—主轴箱;2—刀架;3—尾座;4—床身;5—进给轴;6—丝杠;7—溜板箱

可用车床生产的物品包括烛台架、枪筒、桌球杆、桌腿、碗、棒球棒、乐器(特别是木管乐器)、曲轴和凸轮轴等。

铣床是用来加工固体材料的机床。铣床按机床主轴的方向常被划分为卧式和立式两种基本的类型。这两种类型铣床的尺寸均有小至适合于安装在工作台上的,也有房间大小的。它不像钻床是把工件固定,由钻头轴向移动来钻进材料,铣床上的工件可以按和铣刀旋转相反的方向径向移动,以完成工件两侧和顶端的铣削。工件和铣刀的移动可以通过精密的载玻片和导螺杠或类似技术来控制,其精度值小于 0.001 in(0.025 cm)。铣床可以手工操作,也可采用机械自动化操作或通过计算机数字控制进行数字自动化操作。铣床可以执行从简单(键槽的铣削、刨平、钻孔)到复杂(如轮廓修整、刻模)的大量操作。切削液由泵抽送到切削点,以便于冷却、润滑和冲洗掉切屑。

牛头刨床是一种由工件和单刃刀具的相对直线运动来完成加工的机床。刨床与车床类似,不过它的切削是直线的而不是螺旋的(增加运动轴也可以产生螺旋路径,做螺旋刨削)。牛头刨床类似于龙门刨床,只是它更小,刀具装在滑枕上,在固定工件的上方运动,而不是整个工件在刀具下方运动。滑枕通过内置的曲柄来回地运动。也有牛头刨床是由液压力来驱动的。

牛头刨床主要分为拉削式、卧式、立式、通用型、齿轮传动型、曲柄传动型、液压传动型、轮廓和移动刀架式,其中,卧式牛头刨床是最常用的。立式牛头刨床一般安装有转盘,以便于加工曲面(和螺旋刨削原理相同)。立式刨床基本上和插床一样,虽然它们有技术上的不同——如果把一个真正的牛头刨床定义为一台滑块可以垂直运动的机器。插床是被固定在竖直平面上的。

手动式小型的牛头刨床已经成功生产出来。随着尺寸的增大,机床的质量和它的动力要求也在增大,因此有必要使用电动机或其他机械动力的供应来满足这些要求。电动机驱动机械装置(由主动轮、从动轮、曲柄或链轮和链条)或由一个液压马达通过液压缸来提供必要的运动。

龙门刨床是一种金加工机床,它通过工件和单刃刀具之间的直线往复运动形成线性刀具路径。它的切削类似于车床,但它的刀具路径不是螺旋形的(添加运动轴的话,它也可以产生螺旋形刀具路径)。龙门刨床类似于牛头刨床,但比牛头刨床更大,它的整个工件在刀具下面的工作台上运动,而不是刀具装在滑枕上,在固定工件的上方运动。通过机械装置如齿轮和齿条机构、导螺杆或液压缸,工作台也可在刀头下方的床身上来回运动。

研磨机通常简称为"磨床",是一种用来研磨的机床,它是用砂轮作为切削刀具的。砂轮表面的每个磨粒通过剪切变形从工件表面切削下碎屑。

磨削通常用来做精加工以获得高的表面质量(例如,低的表面粗糙度)与高精度的形状和尺寸。经过磨削,尺寸精度可以达到接近0.000025 mm,因此在大多数情况下,它容易用来做精加工,切削量比较小,为0.25～0.50 mm深。然而,磨床还可用于粗加工,以快速地磨去大量金属。因此磨削可应用于很多领域。

第23课 节能装备

近些年,工业的发展和人口的增长使得全球对能源的需求增量巨大。今天,对能源的需求量几乎是30年前的2倍。根据国际能源机构(IEA)的评估,到2030年,可能再增长50%以上。因此,能源供应远远满足不了实际需求。由于潜在的能源危机近在眼前,全世界必须立刻行动起来,有力地提升能源效率。

在工厂中,节约能源的目标实际上是既要实现生产率的提高又要节能。工厂采用了不同形式的行动来节约能源。

对于空调设施,除了在工厂和建筑里使用更好的热绝缘技术外,使用自然通风装置、太阳能、地热也是人们所希望的。

对于照明,更多地使用自然光以及发光二极管来照明,尤其是用于普通照明的白色发光二极管装置的改进,被视为节约能源的行动。与大部分其他照明灯具相比,白光LED灯拥有较长的预期寿命和更高的效率(同样的光亮使用更少的电)。因为LED灯的尺寸小,照明的空间分布控制极其灵活,并且可以通过控制光线的输出和空间分布排列达到没有功效损失的目的。

此外,低损耗、高效率的电力变压器和电动机在工业领域日益受到重视。电力变压器损耗的分类如图23.1所示。工厂里的变压器全天候工作,而又很少达到额定满负荷的状态。因此,最为重要的一点是要减少在铁芯中由磁滞现象和涡流损失引起的空载损耗。据报道,在载荷系数为28%和50%时,分别采用超级非晶和油浸式变压器,可以使损耗分别减少到50%和33%。

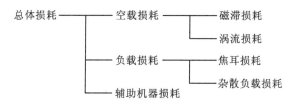

图23.1 电力变压器损耗的分类

对工业和公共事业来说,要降低能量消耗、减少费用支出,最快捷的方法之一是采用高效率的电动机。事实上在每个部门中,采用了成千上万的电动机以驱动机械、压缩机、风扇、泵或者带式运输机等,其用电量约占所有工业用电的67%。人们采用合适的变换器与电动机结合使用来控制电动机的运转速度。用较小的变速电动机来替代普通的电动机,在输出与负载相当的情况下可以节约电能,避免污染,在经济上获益。如果是一台伺服电动机,转子的位子可以被连续地监测和控制,旋转的速度和方向能够迅速、平稳地改变。在伺服电动机系统中,只由旋转运动才会消耗能量。此外,当系统处于待命状态或空载时,就不需要能量了,除非需要保持运动。因此,采用伺服电动机系统能实质上减少能源消耗,从而节约能源。

另一方面,下列技术被应用在各种各样的工业机器和一些设备中以节约能源。大约半个世纪以前,为火车制动开发了一种动力制动方式,通过机车牵引电动机,运动着的火车的动能被用来产生电流。现在,再生制动系统被用在蓄电池或电容器中以存储电能,这些电能是由机器在停转或减速操作中的动能转化而来的。据报道,货运电梯向下的势能可以被转换成再生电能储存在Ni-H蓄电池中,因而减少电梯20%以上的能量消耗。

然而,最好的工业能源节约是对工业过程本身的效率予以改进,例如采用钢的连续铸造,以及就像美国的纸张、板材、胶合板制造业所做的,利用废品来产生电能和热能。

相对于常规的发电厂30%~50%的燃料利用率,利用发电产生的废热为工业或辖区供热,将90%输入的燃料转换成有用的能量,能节约大量的燃料,避免污染。而一些制造出大量的高温液流或废蒸汽的生产设备已经利用废热来进行电力生产。当热能可以就地或近距离使用时,热电联产(CHP)是最有效率的。当热能必须经较长的距离运送时,总体的效率就会降低。这就需要非常隔热的管道,成本高昂且效率低下。然而,在能量损失相同的情况下,电力可以沿着相对简单的电线输送远得多的距离。

2004年,大约81 GW(810亿瓦特)的热电联产设备在美国装机,能提供所生产的全部电力的约12%。欧洲在热电联产装机量上远远领先于美国,北欧国家装机量超过了30%,热电联产还被广泛地用于中国香港以及丹麦、瑞典、荷兰、德国的气候战略中。热电联产的发展潜力还很巨大,例如,美国化学工业只利用了大约30%的热电联产潜能。所有的美国常规发电厂加起来也只将1/3的燃料转化成电能,即有2/3作为废热浪费了,相当于日本所消耗的总能量。完全地采用这一新技术是有利的,能减少全美国23%的二氧化碳排放量。将来自工业过程中的废热卖到距离可以承受的其他地方,从成本效益来说,可以节约美国工业能源的30%。

第24课　生产控制与质量保证

简言之,生产控制是加工中保持适当标准的生产过程的总称。它包括厂家认为为保证一种产品某个特定性能必需的所有控制和检测行为。生产控制的根本目标就是要确保产品或加工过程符合特定的要求,可靠并令人满意。生产控制着重于以下三个方面。

- 各种基础要素,如控制、岗位管理、明确和管理良好的工艺、性能和完整的标准,以及检验记录。
- 技能,如知识、技术、经验以及资质。
- 柔性要素,如个人诚信、自信、组织文化、激励、团队精神以及良好的人际关系。

生产控制包括成品检验,在成品检验中要检验每个产品的外观,并且在产品在市场上出售之前通常要使用立体显微镜做细致入微的检查。检测人员要提供检查清单以及不可接受的产品缺陷的描述,比如裂缝或是表面瑕疵。如果在以上三个方面存在任何一种方式的缺陷,都将危及产品质量。

产品质量控制强调产品测试以发现缺陷并向决定产品发售与否的管理者报告,而质量保证的目的是提高和稳定质量(结合工艺),从而在一开始就避免,至少最小化导致缺陷的问题。那么,何谓质量保证呢?质量保证是指在质量体系中执行的有计划的、系统的活动,用以保证产品的质量要求或服务能够满足要求。它是系统化的措施,是与标准做对照的监测过程,并且构成反馈循环以防患于未然。这一点与质量控制不同,质量控制更侧重于产品本身。由图24.1可发现它们之间的不同。

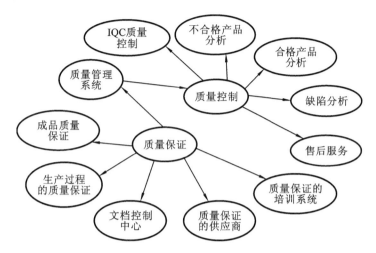

图 24.1 质量控制和质量保证结构示意图

质量保证包含两个原则。"符合目标",即产品应该符合预期目标;"首次就做好",即错误应该被消除。质量保证包括对原材料质量、装配、产品及其零部件、与生产相关的服务、管理,以及生产及检测过程的管理措施。

一般而言,合格的质量由产品的使用者、消费者或顾客认定,而不是由社会决定。它和成本无关,使用像"优""劣"这样的形容词或描述语是不合适的。例如:一种价格低廉的产品,由于它是一次性使用的,顾客可能认为它是高质量的;而另一种价格低廉的产品,由于它不是一次性使用的,可能被顾客认为是质量低劣的产品。

质量保证过程有许多的形式,其范围和深度各不相同。通常为产品的生产过程制订特殊的质量保证过程。

一个典型的质量保证过程包括:
- 以前产品的测试;
- 改进计划;
- 包括改进和调整的设计;
- 生产改进;
- 新项目和改进的审查;
- 新项目的测试。

质量保证的发展经历了三个阶段。

1) 最初的产品质量控制

在中世纪,同业公会对其成员产品的质量控制负责,为其成员的产品设立了一定的标准,并一直采用。

皇家政府像消费者一样非常关注质量控制。由于这个原因,英国国王约翰指派威廉·鲁特姆对战舰的建造和维修作汇报。几个世纪后,英国海军大臣塞缪尔·佩皮斯,指派了更多的类似监工者。在由于工业革命导致的大规模的劳动分工和机械化之前,对工人来说控制其产品的质量是可能的。工业革命推动了一个体系的产生,在这个体系内,大量的人被组织在一起从事一种专门的工作,并且由被指派的工头管理,从而实现对生产过程的质量控制。

2) 战时生产

在第一次世界大战期间，随着更多的工人被管理，生产过程通常变得更加复杂。在这一时期，批量生产和计件生产被引入，工人因为超额生产收入增加，这反过来偶尔又导致这样的问题产生：质量低劣的零件被输送到装配线上。为了应对产品质量低劣的问题，出现了专职的检测人员，以识别、检验产品质量问题并给出解决产品质量问题的方法。二十世纪二三十年代出现的通过检测控制产品质量的做法推动了产品质量检测职能的发展，使其从生产中被分离出来，成为一个独立的部门。

在工业制造领域，对质量的系统研究始于20世纪30年代，主要是在美国，当时废品和返工导致的成本引起了人们的重视。由于第二次世界大战期间大规模生产带来的影响，必须要引入某种能促进对质量控制的形式，即所谓的统计质量控制（SQC）。对SQC早期的研究工作始于1924年的贝尔实验室的沃尔特·休哈特的著名的单页备忘录。

SQC观点认为，并不能将每一件产品进行完全的检测以纳入被接受和不被接受的组别。它通过延长检测阶段和提高检测机构的效率，提供给检测者诸如采样和控制图表等控制工具，但也不可能实现百分之百的检测。标准统计技术允许生产者对一定比例的产品进行质量抽样检测，以使整个批次的质量和生产运行情况达到期望的水平。

3) 战后

在第二次世界大战后的时期，许多国家在战争中被摧毁的制造能力被重新建立起来。道格拉斯·麦克阿瑟将军监督着日本的重建。在这一时期，麦克阿瑟将军对现代质量观念的形成受到两个关键人物——W·爱德华兹·戴明和约瑟夫·朱兰的影响。这两位人物提升了日本的商业和技术团体的质量协作观念，并且这些团体在日本的经济重建中利用了这些观念。

虽然有许多人试图在美国的各行业引入更全面的质量控制方法，但美国仍然采用检测和抽样检验的质量控制理念来消除生产线上的缺陷产品，基本上忽视了质量保证技术几十年来取得的进步。

第25课 农业设备

农业设备是指应用于农业领域或农场里面的设备。它有一段很长的发展史且涵盖的范围很广。

第一个把狩猎和采集的生活方式转变为以种田为生的人可能只能徒手劳作，或许还用一些棍棒或石块作为工具。后来人类发明了刀、镰刀和木犁等劳动工具，并且使用达数千年之久。

随着工业革命的到来和更加先进的机器的发明，农业生产方式产生了巨大的飞跃。利用轮式机械代替锋利的手工工具收割谷物，可以极大地提高效率。利用脱粒机械代替棍子敲打的粮食脱粒方式，可以把谷子从谷穗和秸秆上分离出来。

农业机械早期靠的是马或其他被驯养的动物提供动力。蒸汽动力发明后出现了轻便发动机及后来的牵引机车，它是一种类似于蒸汽机车的地面爬行式多用途的移动动力源。早期的牵引机车如图25.1所示。农业蒸汽机接过了马匹沉重的拖拉的工作。蒸汽机上通常还装有一个带轮，带轮可以通过长的传动带给固定安装的机械提供动力。蒸汽驱动的机械以今天的标准来看效率实在是太低了，但是由于其大尺寸和低齿速比，它们可以提供一个很大的牵引力。它们的低速使得农民认为牵引机只有两种速度：慢速和超慢速。

(a) 1882哈里森机械厂蒸汽拖拉机

(b) 一个非常古老的手动汽油驱动拖拉机

图 25.1　早期牵引动力拖拉机的外形

内燃机，首先是汽油机，随后是柴油机，成为新一代牵引机的主要动力源。这些发动机也推动了自行式联合收割机和脱粒机以及牵引式联合收割机的发展。这些联合收割机在田间不停运动，完成谷物的收割、脱粒，并将谷物与秸秆分开，而不是割下谷物的秸秆，运向一个固定放置的打谷机。

农业设备有很广的范围，在本文中我们将给大家介绍其中的一些。

农用拖拉机用来拉或推农业机械或为犁地、耕种、耙地、播种和类似的任务提供牵引动力。为了实现某种特定用途，已经设计出了各种各样的农用拖拉机。它们包括：中耕作物拖拉机，这种拖拉机可以调整轮胎宽度从而顺利通过玉米、西红柿或其他作物的行间，而不压坏作物；麦地式或标准式拖拉机，其具有不可调整的固定式轮子和较低的重心，以便于耕作和适应作物播种中的各种繁重的野外工作；"高作物"拖拉机，它可以调整轮距和增加离地高度，通常用于棉花和其他长得较高的中耕作物的种植；多用途拖拉机，它是一种具有低重心和小转弯半径的小型拖拉机，用于与农场及其建筑物相关的一般用途。图 25.2 所示为一些现代拖拉机。

(a) 一个大型的、现代化的约翰·迪尔9400型四轮驱动拖拉机

(b) 北达科他州一个现代的可操纵的全履带动力装置在种植小麦

图 25.2　现代的拖拉机

中耕机是复耕时使用农具的一种。中耕机中的一种是带齿（也称为钩爪）的构架，当它在牵引力作用下沿直线运动时将耕作土壤。另一种中耕机是利用圆盘或牙齿的旋转运动实现类似的效果的机器。旋耕机是一种主要的中耕机。中耕机能粉碎并搅拌土壤，可在播种之前使用，以使土壤中饱含空气并平整疏松苗床；或者在作物已经开始生长后使用，以除掉杂草——通过翻耕靠近作物植株的表层土壤把杂草的根挖除从而杀死植株周围的杂草，或者把杂草的叶子埋入表土中来干扰杂草的光合作用，或者两种方法一起使用。中耕与耙地不同。耙地是翻耕整个的表层土壤，而中耕是非常精细地翻耕土壤，在保护作物植株的同时翻耕杂草。中耕

机通常要么是自行式的,要么被设计成一个附件被安装在两轮或四轮拖拉机后面。如果安装在两轮拖拉机后面,中耕机通常被刚性固定并通过联轴器得到拖拉机输送的动力。如果安装在四轮拖拉机后面,中耕机通常通过一个三点式悬挂装置连接在拖拉机上并用一个动力输送装置驱动。牵引杆式连接在全世界范围依然普遍使用。现在仍然有时牲畜作为动力,这种情况在发展中国家较为常见,而在较发达的工业化经济体中非常罕见。

犁是农事中对土壤进行整治以为播种或栽植作准备的工具或机器。在大部分有记载的历史中,它是农业生产中的基本工具,并代表农业所取得的重大发展之一。由图25.3中早期和现代的犁,我们可以了解犁的发展。

(a) Chinese iron plow　　　(b) A four-furrow reversible plough

图 25.3　早期的和现代化的犁

翻耕的主要目的是翻转上层土壤,将新鲜的养分带到表面,并且掩埋杂草和之前作物的残留部分以及杂草的种子,不让这些种子发芽。翻耕还可以使土壤饱含空气,使得土壤能保持更多的水分从而为种植新的作物提供更好的培养基。在现代耕作过程中,通常会让翻耕过的田晒干,并且在栽植作物之前再耙一遍。为耕作爱好者举办的现代赛事比如"全国翻耕锦标赛"中犁最开始是使用牛作为牵引动力的,后来在许多地方也使用马(一般是役马)或骡子。在工业化中,首先用来驱动犁的机械是蒸汽驱动的牵引发动力或蒸汽拖拉机,但这些动力装置后来都逐渐被内燃机驱动的拖拉机所取代。在过去的二十年中,有些地区由于土壤破坏和侵蚀问题,已经减少了犁的使用,而浅耕和其他减少土壤侵蚀的耕作技术正受到青睐。

撒种机,有时也称为农业播撒机,是农业中常见的一种工具,通常用来撒布种子、石灰、肥料、沙子、融雪剂等,是下落式播种机的替代产品。

撒肥机又叫牲畜粪摊铺机,是一种把粪便当做肥料撒布在田里的农用机械。典型的现代撒肥机挂拖在拖拉机后面,并且带有一个由拖拉机动力输出装置驱动的旋转机构。在北美,车载式撒肥机也很常见。

滴管灌溉,也称为涓流灌溉或微灌,或局部灌溉,是灌溉方式的一种。采用这种灌溉方式时,通过一个带有阀门、支管、毛管、滴头的网络使水慢慢地滴在位于土壤表层的根部或直达作物根系区域,从而节水省肥。它能在细小软管帮助下将水直接输送到植株的底部。

联合(或简单联合)收割机是一种收割粮食的机器。典型的联合收割机如图25.4所示。"联合收割机"这个名字来源于实践:它能将三种不同的操作,即收割、脱粒和扬谷,组合到一个单一的过程中。能用联合收割机收割的农作物包括小麦、燕麦、黑麦、大麦、玉米、大豆和亚麻(亚麻籽)。收割后面遗留在田里的秸秆是作物剩下的干缩的杆茎和叶子,它们还有有限的养分,通常要么剁碎了撒在田里,要么捆扎起来,作为牲畜食用的草料或是垫草。

(a) 一种典型的联合收割机　　(b) 约翰迪尔泰坦系列联合卸荷玉米收割机

图 25.4　联合收割机

联合收割机是最重要的节省人力的发明之一,使得从事农业生产的人口只占到人口总数的一小部分。

在上个世纪,农业机械的主要技术基本变化不大。虽然现代化的收割机和播种机可以把工作完成得更好或者比其"前任"的工作有改进,但是现在价格 25 万美元的联合收割机所进行的仍然只是收割、脱粒和谷粒分离这些工作,这些工作在本质上与以前所做的没什么两样。然而,技术正在改变人类操作机械的方式。计算机监控系统、GPS 定位器和自行引导程序的使用,可以让最先进的拖拉机和机具在使用燃料、种子和化肥的时候更加精确和减少浪费。在可预见的将来,一些农业机械将能够通过使用 GPS 地图和电子传感器而具备自动驾驶能力。更为神秘的是纳米技术和基因工程等新领域。在这些领域微观的设备和生物过程,被分别以不同寻常的新方式用在完成农业的新机器上。

农业可能是最古老的行业之一,但机械的发展和使用已经使得愿意从事农业的人非常的稀少。

并不是每个人都要努力工作,为他们自己生产食物,现在,在美国只有不到百分之二的人口从事着农业生产。但正是这百分之二的人口提供的食物,已经远远超过了其他百分之九十八的人所需要的数量。据估计,20 世纪初,一个美国农民可以养活 25 个人,但今天,这个比例是 1∶130,甚至在一个现代化的农场,一个农民生产的谷物可以养活上千人。随着农业机械化程度的不断提高,农民将会变得越来越专业化和罕见。

第 26 课　质量和环境管理体系

消费者对高质量产品、低服务价格的需求在不断增加,而且他们希望供应商能够持续、可靠地提供优质产品和服务。因此,创立国际认同并且遵守的关于质量控制产品可靠性和安全性的方法已成为必要。此外,建立关注环境和生活质量问题的新的国际标准同样重要。

1. ISO 9000 质量管理标准

质量管理体系(QMS)可以解释为实施质量管理所需要的组织结构、程序、过程和资源。早期的质量管理体系强调根据简单的数据和随机取样对工业产品生产线进行预测。进入 20 世纪以后,由于在大多数工业企业中对劳动力的投入成为最昂贵的投入,因此质量管理的重点转到对团队合作和生产过程动态的管理上,尤其是通过不断的改进在早期就发现出现问题的信号。1987 年国际标准化组织(ISO)颁布了第一套质量管理标准 ISO 9000。ISO 9000 标准系列由一套标准体系和公司认证程序组成。它适用于所有类型的公司,并且已得到全球性的认同,对制造公司的国际贸易方式产生了深远的影响。通过接受 ISO 9000 认证,企业证实自己达到了国际标准化组织规定的标准。在许多行业中,ISO 认证已经成为开展贸易的基本要求。2000 年 12 月,国际标准化组织对 ISO 9000 进行了第一次修订,引进了以下三条新标准。

• ISO 9000:2000——质量管理体系基础和规范:给出了标准中使用的术语和定义,这些术语和定义是正确理解标准体系的基础。

• ISO 9001:2000——质量管理体系要求:作为组织质量管理体系认证的标准,用来证明组织质量管理与满足消费者需求的一致性。

• ISO 9004:2000——质量管理体系业绩改进指南:为建立质量管理体系提供指南,其重点不仅在于要满足顾客的需求,而且要求持续改进。

ISO 9000 标准通过八项原则强调以客户为为中心、领导作用、全员参与、过程方法、系统管理方法、持续改进、准确记录和分析数据、改善与客户的关系的重要性。

原则 1——以客户为中心的组织:组织依存于其客户,故应理解客户当前和未来的需要,满足客户的要求并努力超过客户的预期。

原则 2——领导作用:领导为组织确立统一的目标及方向。他们应建立和维护内部环境,使所有员工充分投入到组织目标的实现中去。

原则 3——全员参与:不同阶层的工作人员都是组织的重要成员,他们充分参与,才能使他们的才干为组织带来利益。

原则 4——过程方法:将相关的资源和行动作为一个过程来管理,会更有效地达到期望的结果。

原则 5——系统管理方法:识别、理解和管理由互相联系的过程构成的、针对预定目标建立的系统,将改善组织的效率和效果。

原则 6——持续改进:持续改进本身应成为组织的永久目标。

原则 7——基于事实的决策方法:有效的决策基于对数据和资料的分析。

原则 8——互利的供需关系:一个组织与其供方之间的关系是互相依存、互惠互利的关系,这种关系能提高双方创造价值的能力。

为了得到 ISO 9000 的认证,企业必须提供大量质量管理过程的文案记录文件,其中包括质量的监控方法、职工培训方法和频率、工作描述、检验程序及使用的统计过程控制工具等。所有过程的高质量记录非常重要。之后由 ISO 9000 审计员来审核,审计员要实际考察企业以确保其有很完整的质量管理体系记录文件并且其质量管理程序符合标准要求。如果审计员发现所有程序都执行有序,则予以认证。一旦公司通过认证,公司将会被登记在 ISO 9000 目录上,这个目录上列出了所有被认证的公司。整个认证过程需花费 18~24 个月,认证需花费 10,000~30,000 美元。每隔三年公司还需要被 ISO 重新审核。ISO 认证的一个缺点是它只关注质量管理程序的运用及是否符合规范,并不涉及产品本身的质量问题以及产品是否满足顾客和市场需求。

2. ISO 14000 环境管理体系

什么是环境管理体系(EMS)? 它是如何运行的? 环境管理体系是实施环境政策和实现环境目标的框架体系。环境管理体系的运行包括记录事实,通过已记录的事实,能向人们展示已完全达到了标准。到目前为止,已有三套环境管理标准体系,其中 ISO 14000 可能是众所周知的一种。

ISO 14000 是由国际组织在 1996 年 9 月制定的一套标准体系。它关注一个组织在其产品生命周期中的活动对环境的影响。组织活动可能是内部的或外部的,涉及从产品生产到报废后处置的全过程,并且包括对环境的影响,例如污染、废弃物的产生与处理、噪声、自然资源的消耗以及能源利用等方面。简单地说,ISO 14000 系列涵盖以下标准。

- ISO 14001——环境管理体系(EMS)。环境管理体系的基本要素包括环境方针、计划、实施、核实、经营调查等。
- ISO 14004——EMS 制定和实施的一般指南。
- ISO 14010-12——环境审核原则和指南。
- ISO 14031——环境绩效评估指南。
- ISO 14020-24——环境标志指南(产品)。
- ISO 14040-45——生命周期评价原则和指南(主要产品)。
- ISO 14050——术语和定义。

ISO 14000 体系包括了组织制定环境管理系统所需要的所有要素,提出了组织为获得第三方注册或认证必须符合的所有要求。

图 26.1 所示为 ISO 14001 环境管理计划框架。首先要建立明确的环境目标(环境方针承诺)。然后,评估企业活动和过程(环境审核),以及运行过程对环境的影响(影响因子分析)。通过上述信息,能够建立环境改进标准(目标和指标)和建立达到目标和指标的程序(运营和管理规程)。当完成上述工作后,可以对在目标和指标方面取得的成绩进行审核(环境管理体系审核),最后根据审核结果(管理评审)对企业所建立的环境体系进行复核。环境管理标准体系强调的是持续改进,一般通过两种方式实现:首先,当企业预定的环境目标实现之后,要确立新的目标使企业持续以更加环境友好的方式运行;第二,环境管理标准体系也在不断简化和完善,例如,最初制定的标准规程因要求企业准备不必要的文字材料而显得过于烦琐。定期召开由组织员工参加的环境管理状况调查会议,对精简事务和保证环境管理工作顺利开展是很有帮助的。

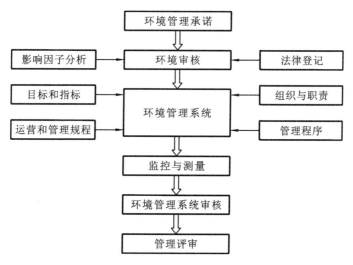

图 26.1　环境管理流程图

ISO 14000 注册意味着正式承认一个组织的环境管理能力符合环境管理标准体系的要求。一个组织会简单地宣称其环境管理体系满足 ISO 14001 的要求(自我声明)。但是,大多数的组织为了获得更多的客户和公众认可或是因为监管机构和客户要求,会选择环境体系认证。目前在许多国家,获得标准认证的公司的数量在迅速增加。

3. ISO 9000 与 ISO 14000 标准体系的关系

ISO 9000 主要关注质量管理。在 ISO 9000 中"质量"的定义是指所有顾客要求的产品或服务的特征。质量管理是组织为确保其产品达到消费者的要求而作出的努力。ISO 9000 体系是专门为满足和达到客户对产品质量的需求和期望而建立的。ISO 14000 主要关注环境管理,环境管理是组织为减少由其活动对环境所造成的有害影响而作出的努力。

尽管 ISO 9000 和 ISO 14000 标准的程序内容差别很大,但是两种标准的框架是非常相似的,例如二者均包含文件记录、管理系统审核、运行控制、记录保持控制、管理政策、审核、培训以及纠正和预防措施。ISO 9000 和 ISO 14000 标准体系的成功推行,要求高层管理者支持和认可,并且要求组织具备一套建立和审查其目标方针(不管其是否与质量或环境有关)的体系。两者都要求组织对管理体系和其目标提供持续的管理审查。如果 ISO 9000 已经推行,引入 ISO 14000 就会相对简单。相反,在没有推行任何质量管理体系的情况下,如果先引进 ISO 14000,也会大大减少公司未来获得 ISO 9000 认证所需要的时间和努力。

参 考 文 献

[1] BEER F P,JOHNSTON J E R. Statics and Mechanics of Materials [M]. New York:McGraw-Hill, Inc. 1992.
[2] MERIAM J L, Kraige L G, HOBOKEN N J. Engineering mechanics [M]. New York:John Wiley & Sons, 2007.
[3] GOC R . "Dynamics" (Physics tutorial). (2004－2005 copyright date)[2010-02-18]. http://www.staff.amu.edu.pl/~romangoc/M3-dynamics.html. Retrieved.
[4] 孟元庆. 力学专业英语[M].哈尔滨:哈尔滨工业大学出版社,2002.
[5] 黄平. 现代设计理论与方法[M]. 北京:清华大学出版社,2010.
[6] NARAYAN K L. Computer Aided Design and Manufacturing [M]. New Delhi:Prentice Hall of India,2008.
[7] MADSEN D A,MADSEN D P, Engineering Drawing & Design [M]. South Melbourne:Cengage Learning,2001.
[8] POTTMANN H, BRELL-COKCAN S, Wallner J. Discrete surfaces for architectural design[M]//Anon. Curve and Surface Design:Avignon 2006. Brentwood :Nashboro Press,2006.
[9] FARIN G . Curves and Surfaces for CAGD:A Practical Guide[M]. London:Academic Press,2002.
[10] MCLENNAN J F. The Philosophy of Sustainable Design [M]. Oregon:Ecotone Publishing,2004.
[11] YANG F S, FREEDMAN B,CTOE R. Principles and practice of ecological design[J]. Environmental Reviews,2004, 12:97-112.
[12] 桂慧.机械专业英语[M].北京:国防工业出版社,2010.
[13] 程安宁,周新建.机械工程科技英语[M].西安:西安电子科技大学出版社,2007.
[14] 程安宁,周新建.机械工程科技英语[M].西安:西安电子科技大学出版社,2007.
[15] 章跃.机械制造专业英语[M].北京:机械工业出版社,2012.
[16] 廖宇兰.机械工程专业英语 [M].北京:化学工业出版社,2009.
[17] 唐一平.机械工程专业英语.[M].北京:电子工业出版社,2009.
[18] 叶邦彦.机械工程专业英语 [M].北京:机械工业出版社,2012.
[19] 黄丽容.机电专业英语 [M].北京:国防工业出版社,2010.
[20] 董华.如何写英文科技论文[J].雁北师范学院学报,2006,22(2):4-6.
[21] 赵运才.机电工程专业英语[M].北京:北京大学出版社,2006.
[22] 谢小苑.科技英语翻译技巧与实践[M].北京:国防工业出版社,2008.
[23] 宋宏.前沿科技英语阅读文选——机电工程篇[M].北京:国防工业出版社,2007.
[24] 中国人民解放军总装备部军事训练教材编辑工作委员会.科技英语翻译实用教程[M].北京:国防工业出版社,1999.

[25] 李光布,饶锡新.机械工程专业英语[M].武汉:华中科技大学出版社,2008.
[26] 宋瑞苓.机电工程专业英语[M].北京:化学工业出版社,2010.
[27] 王建武,李民权,曾小珊.科技英语写作(理论·技巧·范例)[M].西安:西北工业大学出版社,2008.
[28] 陈统坚.机械工程英语[M].北京:机械工业出版社,1995.
[29] 卜玉坤.大学专业英语－机械英语[M].北京:外语教学与研究出版社,2001.
[30] 赵萱,郑仰成.科技英语翻译[M].北京:外语教学与研究出版社,2006.
[31] 上海市职业技术教育课程改革与教材建设委员会.机电与数控专业英语[M].北京:机械工业出版社,2007.
[32] 姜少杰,王永鼎.机电工程专业英语[M].上海:同济大学出版社,2006.
[33] Goddard Space Flight Center. Drawing Standards Manual[M]. Maryland: s. n. , 1994.
[34] KAZMIERCZYK P. Manual on the development of cleaner production policies approaches and instruments[M]. Vienna: Unido CP Programme, 2002.
[35] THOEPE B. Clean Production Strategies: What is Clean Production [EB/OL]. http://cleanproduction. org/library/Factsheet1_Clean_Production. pdf.
[36] TINSLEY S. Environmental Management Plans Demystified: A guide to implementing ISO 14001[M]. London: Spon Press , 2001.
[37] TCHOBANOGLOUS G, THEISEN H, VIGIL S. Intergrated solid waste management engineering principles and management issues[M]. The McGraw-Hill Companies, Inc. 2000.
[38] LEKANG O. Aquaculture Engneering[M]. UK: Blackwell Publishing Ltd, 2007.
[39] CONSTABLE G, SOMERVILLE B. A Century of Innovation: Twenty Engineering Achievements That Transformed Our Lives[M]. Washington, DC: Joseph Henry Press,2003.
[40] JURAN J A, GODFREY A B. Juran's Quality Handbook[M]. New York: McGraw Hill, 1999.
[41] PYZDEK T, KELLER P A. Quality Engineering Handbook[M]. New York: Marcel Dekker Incorporated, 2003.
[42] 施平.机械工程专业英语教程[M].北京:电子工业出版社,2008.
[43] BURR A H, CHEATHAM J B. Mechanical Analysis and Design [M]. London: Prentice Hall,1995.
[44] ASHBY M F. Materials Selection in Mechanical Design[M]. London: Butterworth-Heinemann, 2010.
[45] NORTON R L. Machine Design: An Intergrated Approach[M]. London: Prentice-Hall, 1996.
[46] HIGLEY J E, MISCHKE S C. Mechanical Engineering Design[M].6版.北京:机械工业出版社,2002.
[47] 王群,童长清.模具专业英语[M].北京:机械工业出版社,2007.